Advances in Intelligent and Soft Computing 74

Editor-in-Chief: J. Kacprzyk

T0135204

Advances in Intelligent and Soft Computing

Editor-in-Chief

Prof. Janusz Kacprzyk
Systems Research Institute
Polish Academy of Sciences
ul. Newelska 6
01-447 Warsaw
Poland
E-mail: kacprzyk@ibspan.waw.pl

Further volumes of this series can be found on our homepage: springer.com

Miguel P. Rocha,
Florentino Fernández Riverola, Hagit Shatkay,
and Juan Manuel Corchado (Eds.)

Advances in Bioinformatics

4th International Workshop on Practical
Applications of Computational Biology
and Bioinformatics 2010 (IWPACBB 2010)

 Springer

Editors

Miguel P. Rocha
Dep. Informática / CCTC
Universidade do Minho
Campus de Gualtar
4710-057 Braga
Portugal

Hagit Shatkay
Computational Biology and
Machine Learning Lab
School of Computing
Queen's University Kingston
Ontario K7L 3N6
Canada
E-mail: shatkay@cs.queensu.ca

Florentino Fernández Riverola
Escuela Superior de
Ingeniería Informática
Edificio Politécnico,
Despacho 408
Campus Universitario
As Lagoas s/n
32004 Ourense
Spain
E-mail: riverola@ei.uvigo.es

Juan Manuel Corchado
Departamento de Informática
y Automática
Facultad de Ciencias
Universidad de Salamanca
Plaza de la Merced S/N
37008 Salamanca
Spain
E-mail: corchado@usal.es

ISBN 978-3-642-13213-1   e-ISBN 978-3-642-13214-8

DOI 10.1007/978-3-642-13214-8

Advances in Intelligent and Soft Computing   ISSN 1867-5662

Library of Congress Control Number: 2010928117

© 2010 Springer-Verlag Berlin Heidelberg

Typeset & Cover Design: Scientific Publishing Services Pvt. Ltd., Chennai, India.

Printed on acid-free paper

5 4 3 2 1 0

springer.com

Preface

The fields of Bioinformatics and Computational Biology have been growing steadily over the last few years boosted by an increasing need for computational techniques that can efficiently handle the huge amounts of data produced by the new experimental techniques in Biology. This calls for new algorithms and approaches from fields such as Data Integration, Statistics, Data Mining, Machine Learning, Optimization, Computer Science and Artificial Intelligence.

Also, new global approaches, such as Systems Biology, have been emerging replacing the reductionist view that dominated biological research in the last decades. Indeed, Biology is more and more a science of information needing tools from the information technology field. The interaction of researchers from different scientific fields is, more than ever, of foremost importance and we hope this event will contribute to this effort.

IWPACBB'10 technical program included a total of 30 papers (26 long papers and 4 short papers) spanning many different sub-fields in Bioinformatics and Computational Biology. Therefore, the technical program of the conference will certainly be diverse, challenging and will promote the interaction among computer scientists, mathematicians, biologists and other researchers.

We would like to thank all the contributing authors, as well as the members of the Program Committee and the Organizing Committee for their hard and highly valuable work. Their work has helped to contribute to the success of the IWAPCBB'10 event. IWPACBB'10 wouldn't exist without your contribution.

Miguel Rocha   Juan Manuel Corchado
Florentino Fernández Riverola   Hagit Shatkay
IWPACBB'10 Organizing Co-chairs   IWPACBB'10 Programme Co-chairs

Organization

General Co-chairs

Miguel Rocha   University of Minho (Portugal)
Florentino Riverola   University of Vigo (Spain)
Juan M. Corchado   University of Salamanca (Spain)
Hagit Shatkay   Queens University, Ontario (Canada)

Program Committee

Juan M. Corchado   University of Salamanca (Spain)
  (Co-chairman)
Alicia Troncoso   Universidad of Pablo de Olavide (Spain)
Alípio Jorge   LIAAD/INESC, Porto LA (Portugal)
Anália Lourenço   University of Minho (Portugal)
Arlindo Oliveira   INESC-ID, Lisboa (Portugal)
Arlo Randall   University of California Irvine (USA)
B. Cristina Pelayo   University of Oviedo (Spain)
Christopher Henry   Argonne National Labs (USA)
Daniel Gayo   University of Oviedo (Spain)
David Posada   Univ. Vigo (Spain)
Emilio S. Corchado   University of Burgos (Spain)
Eugénio C. Ferreira   IBB/CEB, University of Minho (Portugal)
Fernando Diaz-Gómez   University of Valladolid (Spain)
Gonzalo Gómez-López   UBio/CNIO, Spanish National Cancer Research
                            Centre (Spain)
Isabel C. Rocha   IBB/CEB, University of Minho (Portugal)
Jesús M. Hernández   University of Salamanca (Spain)
Jorge Vieira   IBMC, Porto (Portugal)
José Adserias   University of Salamanca (Spain)
José L. López   University of Salamanca (Spain)
José Luís Oliveira   Univ. Aveiro (Portugal)
Juan M. Cueva   University of Oviedo (Spain)
Júlio R. Banga   IIM/CSIC, Vigo (Spain)

Kaustubh Raosaheb Patil	Max-Planck Institute for Informatics(Germany)
Kiran R. Patil	Biocentrum, DTU (Denmark)
Lourdes Borrajo	University of Vigo (Spain)
Luis M. Rocha	Indiana University (USA)
Manuel J. Maña López	University of Huelva (Spain)
Margarida Casal	University of Minho (Portugal)
Maria J. Ramos	FCUP, University of Porto (Portugal)
Martin Krallinger	CNB, Madrid (Spain)
Nicholas Luscombe	EBI (UK)
Nuno Fonseca	CRACS/INESC, Porto (Portugal)
Oscar Sanjuan	University of Oviedo (Spain)
Paulo Azevedo	University of Minho (Portugal)
Paulino Gómez-Puertas	University Autónoma de Madrid (Spain)
Pierre Balde	University of California Irvine (USA)
Rui Camacho	LIACC/FEUP, University of Porto (Portugal)
Rui Brito	University of Coimbra (Portugal)
Rui C. Mendes	CCTC, University of Minho (Portugal)
Sara Madeira	IST/INESC, Lisboa (Portugal)
Ségio Deusdado	IP Bragança (Portugal)
Vítor Costa	University of Porto (Portugal)

Organizing Committee

Miguel Rocha (Co-chairman)	CCTC, Univ. Minho (Portugal)
Florentino Fernández Riverola (Co-chairman)	University of Vigo (Spain)
Juan F. De Paz	University of Salamanca (Spain)
Daniel Glez-Peña	University of Vigo (Spain)
José P. Pinto	University of Minho (Portugal)
Rafael Carreira	University of Minho (Portugal)
Simão Soares	University of Minho (Portugal)
Paulo Vilaça	University of Minho (Portugal)
Hugo Costa	University of Minho (Portugal)
Paulo Maia	University of Minho (Portugal)
Pedro Evangelista	University of Minho (Portugal)
Óscar Dias	University of Minho (Portugal)

Contents

Microarrays

Data Mining and Data Integration

Phylogenetics and Sequence Analysis

Biomedical Applications

Bioinformatics Applications

Highlighting Differential Gene Expression between Two Condition Microarrays through Heterogeneous Genomic Data: Application to *Lesihmania infantum* Stages Comparison

Liliana López Kleine and Víctor Andrés Vera Ruiz

Abstract. Classical methods for the detection of gene expression differences between two microarray conditions often fail to detect interesting and important differences, because they are weak in comparison with the overall variability. Therefore, methodologies that highlight weak differences are needed. Here, we propose a method that allows the fusion of other genomic data with microarray data and show, through an example on *L. infantum* microarrays comparing promastigote and amastigote stages, that differences between the two microarray conditions are highlighted. The method is flexible and can be applied to any organism for which microarray and other genomic data is available.

1 Introduction

Protozoan of the genus *Leishmania* are parasites that are transmitted by blood-feeding insect vectors to mammalian hosts, and cause a number of important human diseases, collectively referred as leishmaniasis. During their life cycle, these parasites alternate between two major morphologically distinct developmental stages. In the digestive tract of the sandfly vector, they exist as extracellular elongated, flagellated, and motile promastigotes that are exposed to pH 7 and fluctuating temperatures averaging 25°C. Upon entry into a mammalian host, they reside in mononuclear phagocytes or macrophages (37ªC), wherein they replicate as circular, aflagellated and non-motile amastigotes. In order to survive, these two extreme environments, *Leishmania sp.* (*L. sp*) has developed regulatory mechanisms that result in important morphological and biochemical adaptations [1, 2, 3].

Liliana López Kleine · Victor Andrés Vera Ruiz
Universidad Nacional de Colombia (Sede Bogotá), Cra 30, calle 45, Statistics Department
e-mail: llopezk@unal.edu.co, vaverar@unal.edu.co

M.P. Rocha et al. (Eds.): IWPACBB 2010, AISC 74, pp. 1–8, 2010.
springerlink.com

Microarray studies allow measuring the expression level of thousands of genes at the same time by just one hybridization experiment and the comparison of two conditions (here the development stages of *L.* Sp.). Several microarray analyses have been done to study global gene expression in developmental stages of *L. sp.* [3, 4, 5]. Results show that *L.* sp genome can be considered to be constitutively expressed, as more than 90% of the genes is expressed in the same amount in both stages, and that only a limited number (7-9.3%) of genes show stage-specific expression [3, 6]. Furthermore, no metabolic pathway or cell function characteristic of any stage has been detected [3, 4, 6, 7, 8, 9]. This is an astonishing result because morphological and physiological differences between the two stages are huge, indicating that different specialized stage proteins are needed and therefore gene expression should change. In order to explain the weak differences in gene expression between both stages, it has been proposed that regulations and adaptations take place after translation and not at the gene expression level [10].

The detection of gene expression differences has been of great interest [11], as an understanding of the adaptation and resistance factors of *Leishmania sp.* can provide interesting therapeutic targets. Analytical methods used until now, even improved ones such as the Gene Set Enrichment Analysis [13], have difficulties in the determination of differences between gene expression in both *L. sp.* life cycle stages because in the context of global gene expression variation, the detection of weak differences is not possible using classical methods. Microarray data analysis can still be improved if a way to highlight weak differences in gene expression between *L. sp.* stages is found.

The present method consists of using additional information to the microarrays to achieve this. It allows incorporating different genomic and post-genomic data (positions of genes on the chromosome, metabolic pathways, phylogenetic profiles, etc.) to detect differences between two experimental microarray conditions The method can be applied to detect gene expression differences between any two conditions and for all completely sequenced organisms if genomic data are available. It will be especially useful for the comparison of conditions in which apparently, using classical methods, gene expression seems small.

To apply the proposed strategy, four steps need to be taken: i) Database construction of the genomic data for the same genes that are present on the microarray, ii) kernel construction and parameter estimation, iii) determination of differences in gene expression between two microarray conditions at the kernel level, and iv) interpretation of differences in regard of the original microarray data.

In the present work, we apply the proposed methodology to determine differences between *L. infantum* amastigotes and promastigotes microarray data obtained by Rochette et al [3]. The methodology is proposed for all genes of a microarray data set. Nevertheless, taking into account the interest in determining adaptation and defense mechanisms in the pathogen, we concentrated on genes that could explain the adaptation and resistance of *L. sp.* to the environmental changes between a promastigote and an amastigote. Therefore, we searched primarily for changes in expression of 180 known and putative transport proteins and stress factors.

2 Methodology

2.1 Data Base Construction of Microarray and Genomic Data

2.1.1 Microarrays

The data used are microarray data from Rochette et al. [3]. From these data, we extracted only the expression data comparing promastigotes and amastigotes of *L. infantum* (8317 genes for 14 replicates). Microarray data was downloaded from the NCBI's GEO Datasets [14] (accession number GSE10407). We worked with normalized and log2 transformed gene expression intensities from the microarray data obtained by Rochette and colleagues [3].

2.1.2 Phylogenetic Profiles

They were constructed using the tool Roundup proposed by DeLuca et al [15] (http://rodeo.med.harvard.edu/tools/roundup) which allows the extraction of the presence or absence of all genes of an organism in other organisms chosen by the user. The result can be retrieved as a phylogenetic profile matrix of presence and absence (0,1) of each *L. sp.* gene in the genomes of the other organisms. We generated a phylogenetic profile matrix for 30 organisms[1] and 2599 genes of *L. infantum* sharing common genes.

[1]*A._thaliana, Bacillus_subtilis, C._elegans, Coxiella_burnetii (Cb), Cb_CbuG_Q212, Cb_Dugway_7E9-12, Cb_RSA_331, D._melanogaster, Enterobacter_638, E._faecalis_V583, Escherichia_coli_536, H._sapiens, Lactobacillus_plantarum, Lactococcus._lactis, Listeria_innocua, L._monocytogenes, Mycobacterium_bovis, M._leprae, Nostoc_sp, P._falciparum, Pseudomonas_putida_KT2440, S._cerevisiae, Salmonella_enterica_Paratypi_ATCC_9150, Staphylococcus_aureus_COL, Staphylococcus_epidermidis_ATCC_12228, Streptococcus_mutans, S._pyogenes_MGAS10394, S._pyogenes_MGAS10750, T._brucei, V._cholerae*

2.1.3 Presence of Genes on Chromosomes

For the same 8317 genes we obtained the presence of each of them on the 36 chromosomes of *L. infantum*. This information was obtained directly from the gene name as registered in NCBI (http://www.ncbi.nlm.nih.gov/) and used to construct a presence and absence (0,1) table for each gene in each chromosome (8317 x 36).

2.1.4 Genes of Interest

We constructed a list of 180 genes coding for known or putative transporters and stress factors annotated with these functions in GeneDB (http://www.genedb.org//genedb/).

Once all data types were obtained, determining genes present in all databases was done automatically using functions written in R [16]. Without taking into account the 180 genes of interest, 2092 genes were common to microarrays, presence on the chromosomes and phylogenetic profiles. Using only the 180 genes of interest, we obtained a list of 161 genes common to all 3 datasets.

2.2 Kernel Construction

We built a kernel similarity matrix for each data type. These representations allow the posterior fusion of heterogeneous data. Data are not represented individually, but through the pair-wise comparison of objects (here 161 genes). The comparison is expressed as the similarity between objects through the data. A comparison function of type $k : X \times X \to R$ is used and the data are represented by a $n \times n$ comparison matrix: $k_{i,j} = k(x_i, x_j)$ [17]. Kernels are semi definite positive matrices, and can be used in several kernel algorithms [17]. There are different ways to construct a kernel. The simplest kernels are linear and Gaussian. They have been used for genomic and post-genomic data in previous works, which aimed to infer biological knowledge by the analysis of heterogeneous data [18].

We built Gaussian kernels for all data-types: $k(x_i, x_j) = e^{-\frac{d(x_i,x_j)}{2\sigma^2}}$,where σ is a parameter and d is an Euclidian distance. The constructed kernels were: K_{A1} for the gene expression data on amastigotes, K_{P1} for the gene expression data on promastigotes, K_2 for the phylogenetic profiles and K_3 for the presence on the chromosomes. The parameters associated to each kernel were $sigma_{1A}$, $sigma_{1P}$, $sigma_2$, $sigma_3$.

Then, we constructed two unique kernels K_{A1sum} and K_{P1sum} for the gene expression data together with the other types of data by addition: $K_{(A,P)1sum} = w_1 K_{(A,P)1} + w_2 K_2 + w_3 K_3$, where w are weights and considered also as parameters.

Taking into account the objectives of the present work (detection of gene expression differences between amastigote and promastigote microarray), parameters sigma and w were found using a search algorithm evaluating all combination of parameters, that optimized the difference between K_{A1sum} and K_{P1sum}. The criterion was the minimum covariance between both kernels. The values tested for σ were: 0.01,0.1,0.5,1,10,15,50 and the values tested for w were 0,0.1,0.5,0.9.

2.3 Detection of Differences between Amastigote and Promastigote Gene Expression

Differences in similarity and gene expression

Kernels K_{A1sum} and K_{P1sum} were compared to detect differences in gene expression between both *L. infantum* stages by computing a *Dif* matrix as follows: *Dif* = K_{A1sum} - K_{P1sum}. The resulting matrix values were ordered, and those with higher value than a certain threshold were retained. The same was done for the kernels without integration of other data types, the ones constructed only based on the microarray data: K_{A1} and K_{P1}. Two thresholds used were: 10% (T1) and 20% (T2) of the maximum distance value found in *Dif*.

Subsequently, a list of pairs of genes implicated in each change of similarity is generated. To interpret this change in similarity we returned to the original microarray data and calculated the sum of gene expression intensities in each condition. This allows to determine which one of the two genes is responsible for the similarity change and finally to identify potential targets that explain the differences that occur for *L. infantum* adaptations during its life cycle. R code is available under request: llopezk@unal.edu.co.

The most interesting targets (which show the highest difference or are present repeatedly on the list), are candidates to perform wet-lab experiments.

3 Results and Discussion

The parameters that were determined by the search algorithm to maximize differences between amastigote and promastigote gene expression are shown in table 1.

Table 1 Parameters obtained optimizing differences between amastigote and promastigote gene expression

Kernel	K_{A1}	K_{P1}	K_2	K_3
Sigma	1	1	15	50
Weight	0.9	0.9	0	0.1

K_2 seemed not to be useful to highlight the differences between gene expressions in both stages. Nevertheless, phylogenetic profiles have shown to be informative in other studies, i.e., for the determination of protein functions [18, 19]. It is possible that either phylogenetic profiles are definitely not useful to highlight the differences between the two conditions analyzed here, or that the organisms that were chosen to construct the profiles are phylogenetically too distant from *L. infantum* and therefore poorly informative. The fact that only a few genes are present in most of them (data not shown), corroborates the second explanation and opens the possibility that phylogenetic profiles could be useful if appropriate organisms were chosen.

The result of the parameter search leaves only the K_3 kernel with a low weight (0.1) to be added to the microarray data kernels. This indicates that apparently the fusion with data other than microarrays is not useful. Nevertheless, the results obtained when similarity changes were detected indicates that including K_3 is indeed useful. The differences detected between K_{A1} and K_{P1} (based only on microarray data) implicate a change in only 10 similarities for threshold 1 (T1) and 44 changes for T2 using the 161×161 kernel. The fusion with K_3 allows the detection of more differences: 14 for T1 and 61 for T2. The 14 gene pairs which show similarity changes for above T1 in the fusion kernels are shown in Table 2. As the behavior of the genes of interest should be regarded in the context of all the genes, the position of these similarity changes when all genes (2092) are analyzed is important. Using these 2092, 1924 similarity changes are detected above T1. The position (Pos) in the 1924 list of genes of interest is also indicated in Table 2 for comparison.

4 Conclusions and Future Work

It is important to point out that the use of genomic data to highlight differences between two microarray conditions is possible and easy to implement via the use of
kernels. The flexibility of kernels allows the fusion of very different types of data,
as only the comparison of objects needs to be computed. Depending on the biological question behind the study, other types of data such as information on Clusters of Orthologous Groups (COGs) or physical interaction between proteins
obtained from two-hybrid data could be included. Graph metabolic pathway information could be very informative and a kernel for graphs has been already
proposed [20].

Differences between two microarray conditions are highlighted by the fusion
with other types of data. Nevertheless, the usefulness of genomic data depends on
their quality and information content. In our example, the phylogenetic profiles
appeared to be useless in highlighting information. Use of more informative
organisms to construct the phylogenetic profiles needs to be investigated.

Table 2 List of 14 gene pair similarity changes between amastigote (K_{A1sum}) and promastigote (K_{P1sum}) microarray data highlighted through the fusion with data on the presence of
genes on chromosomes (K_3). AMA and PRO: sum of gene expression of each gene in the
amasigote microarray (AMA) or promastigote microarray (PRO). Pos: position of similarity
changes in the 2092×2092 kernel (1924 similarity changes above T1). P: gene annotated as
putative. Hpc: hypothetical conserved protein

Gene pair with similarity change in K_{A1sum} vs. K_{P1sum}				Pos	AMA	PRO	AMA	PRO
Gene1		Gene2			Gene1	Gene1	Gene2	Gene2
LinJ07.0700	vacuolar-type Ca2+ATPase, P	LinJ34.2780	Hpc	16	-1,68	-1,84	-3,42	-3,57
LinJ08.1020	stress-induced sti1	LinJ35.2730	Hpc	35	-3,98	-3,18	1,98	4,22
LinJ09.0960	ef-hand prot. 5, P	LinJ23.0290	multidrug resist. P	86	-1,67	-2,36	-2,47	-2,57
LinJ14.0270		LinJ23.0430	ABC trans.-like	287	-3,65	-2,55	-0,29	-0,98
LinJ14.0270	Hpc	LinJ24.1180	Hpc	459	-3,65	-2,55	20,31	26,54
LinJ14.0270	Hpc	LinJ25.1000	Hpc	569	-3,65	-2,55	-4,82	-4,71
LinJ14.1040	Hpc	LinJ33.0340	ATP-bin. P	752	-1,91	-1,01	4,37	6,27
LinJ17.0420	Hpc	LinJ35.2730	Hpc	769	-4,28	-4,41	1,98	4,22
LinJ20.1230	calpain-like cyste-ine pept., P (clcp)	LinJ28.2230	Hpc	865	-3,97	-3,87	-0,06	-0,6
LinJ20.1250	Clcp	LinJ35.2730	Hpc	923	-4,12	-3,98	1,98	4,22
LinJ21.1900	calcineurin B subunit, P	LinJ35.2730	Hpc	965	-4,01	-3,87	1,98	4,22
LinJ22.0050	Hpc	LinJ35.2730	Hpc	1002	-3,43	-3,35	1,98	4,22
LinJ31.1790	Hpc	LinJ35.2730	Hpc	1036	-4,39	-4,12	1,98	4,22
LinJ35.2730	Hpc	LinJ33.2130	Clcp	1782	1,98	4,22	-3,95	-3,83

The present work opens the possibility of implementing a kernel method that will allow determining differences in a more precise way once the data are fused. The detection of differences can be improved in several ways. Here, only a preliminary and very simple comparison of similarities is proposed. The kernel method could be based on multidimensional scaling via the mapping of both kernels on a common space that could allow the measure of distances between similarities on that space.

Although differences between kernels are ordered, having a probability associated to each difference would be useful. This could be achieved by a bootstrapping procedure or a matrix permutation test.

References

[1] McConville, M.J., Turco, S.J., Ferguson, M.A.J., Saks, D.L.: Developmental modification of lipophosphoglycan during the differentiation of *Leishmania major* promastigotes to an infectious stage. EMBO J. 11, 3593–3600 (1992)

[2] Zilberstein, D., Shapira, M.: The role of pH and temperature in the development of *Leishmania* parasites. Annu. Rev. Microbiol. 48, 449–470 (1994)

[3] Rochette, A., Raymond, F., Ubeda, J.M., Smith, M., Messier, N., Boisvert, S., Rigault, P., Corbeil, J., Ouellette, M., Papadopoulou, B.: Genome-wide gene expression profiling analysis of *Leishmania major* and *Leishmania infantum* developmental stages reveals substantial differences between the two species. BMC Genomics 9, 255–280 (2008)

[4] Cohen-Freue, G., Holzer, T.R., Forney, J.D., McMaster, W.R.: Global gene expression in *Leishmania*. Int. J. Parasitol. 37, 1077–1086 (2007)

[5] Leifso, K., Cohen-Freue, G., Dogra, N., Murray, A., McMaster, W.R.: Genomic and proteomic expression analysis of *Leishmania* promastigote and amastigote life stages: the *Leishmania* genome is constitutively expressed. Mol. Biochem. Parasitol. 152, 35–46 (2007)

[6] Saxena, A., Lahav, T., Holland, N., Aggarwal, G., Anupama, A., Huang, Y., Volpin, H., Myler, P.J., Zilberstein, D.: Analysis of the *Leishmania donovani* transcriptome reveals an ordered progression of transient and permanent changes in gene expression during differentiation. Mol. Biochem. Parasitol 52, 53–65 (2007)

[7] Ivens, A.C., Lewis, S.M., Bagherzadeh, A.: A physical map of *Leishmania* major friedlin genome. Genome Res. 8, 135–145 (1998)

[8] Holzer, T.R., McMaster, W.R., Forney, J.D.: Expression profiling by whole-genome interspecies microarray hybridization reveals differential gene expression in procyclic promastigotes, lesion-derived amastigotes, and axenic amastigotes in *Leishmania mexicana*. Mol. Biochem. Parasitol 146, 198–218 (2006)

[9] McNicoll, F., Drummelsmith, J., Müller, M., Madore, E., Boilard, N., Ouellette, M., Papadopoulou, B.: A combined proteomic and transcriptomic approach to the study of stage differentiation in *Leishmania infantum*. Proteomics 6, 3567–3581 (2006)

[10] Rosenzweig, D., Smith, D., Opperdoes, F., Stern, S., Olafson, R.W., Zilberstein, D.: Retooling*Leishmania* metabolism: from sand fly gut to human macrophage. FASEB J. (2007), doi:10.1096/fj.07-9254com

[11] Lynn, M.A., McMaster, W.R.: *Leishmania*: conserved evolution-diverse diseases. Trends Parasitol 24, 103–105 (2008)

[12] Storey, J.D., Tibshirani, R.: Statistical significance for genome-wide experiments. Proc. Natl. Acad. Sci. 100, 9440–9445 (2003)

[13] Subramanian, A., Tamayo, P., Mootha, V.K., Mukherjee, S., Ebert, B.L., Gillette, M.A., Paulovich, A., Pomeroy, S.L., Golub, T.R., Lander, E.S., Mesirov, J.P.: Gene set enrichment analysis: A knowledge-based approach for interpreting genome-wide expression profiles. PNAS 102, 15545–15550 (2005)

[14] Edgar, R., Domrachev, M., Lash, A.E.: Gene Expression Omnibus: NCBI gene expression and hybridization array data repository. Nucleic Acid Res. 30, 207–210 (2002)

[15] DeLuca, T.F., Wu, I.H., Pu, J., Monaghan, T., Peshkin, L., Singh, S., Wall, D.P.: Roundup: a multi-genome repository of orthologs and evolutionary distances. Bioinformatics 22, 2044–2046 (2006)

[16] R Development Core Team R: A language and environment for statistical computing. R Foundation for Statistical Computing. Vienna, Austria (2005), ISBN 3-900051-07-0, http://www.R-project.org

[17] Vert, J., Tsuda, K., Schölkopf, B.: A primer on kernels. In: Schölkopf, B., Tsuda, K., Vert, J. (eds.) Kernel methods in computational biology. The MIT Press, Cambridge (2004)

[18] Yamanishi, Y., Vert, J.P., Nakaya, A., Kaneisha, M.: Extraction of correlated clusters from multiple genomic data by generalized kernel canonical correlation analysis. Bioinformatics 19, 323–330 (2003)

[19] López Kleine, L., Monnet, V., Pechoux, C., Trubuil, A.: Role of bacterial peptidase F inferred by statistical analysis and further experimental validation. HFSP J. 2, 29–41 (2008)

[20] Kondor, R.I., Lafferty, J.: Diffusion kernels on graphs and other discrete structures. In: Sammut, C., Hoffmann, A.G. (eds.) Machine learning: proceedings of the 19th international conference. Morgan Kaufmann, San Francisco (2002)

An Experimental Evaluation of a Novel Stochastic Method for Iterative Class Discovery on Real Microarray Datasets

Héctor Gómez, Daniel Glez-Peña, Miguel Reboiro-Jato, Reyes Pavón,
Fernando Díaz, and Florentino Fdez-Riverola

Abstract. Within a gene expression matrix, there are usually several particular macroscopic phenotypes of samples related to some diseases or drug effects, such as diseased samples, normal samples or drug treated samples. The goal of sample-based clustering is to find the phenotype structures of these samples. A novel method for automatically discovering clusters of samples which are coherent from a genetic point of view is evaluated on publicly available datasets. Each possible cluster is characterized by a fuzzy pattern which maintains a fuzzy discretization of relevant gene expression values. Possible clusters are randomly constructed and iteratively refined by following a probabilistic search and an optimization schema.

Keywords: microarray data, fuzzy discretization, gene selection, fuzzy pattern, class discovery, simulated annealing.

1 Introduction

Following the advent of high-throughput microarray technology it is now possible to simultaneously monitor the expression levels of thousands of genes during important biological processes and across collections of related samples. In this

Héctor Gómez · Daniel Glez-Peña · Miguel Reboiro-Jato · Reyes Pavón
Florentino Fdez-Riverola
ESEI: Escuela Superior de Ingeniería Informática, University of Vigo,
Edificio Politécnico, Campus Universitario As Lagoas s/n, 32004, Ourense, Spain
e-mail: hector.j.gomez@gmail.com,
        {dgpena, mrjato, pavon, riverola}@uvigo.es

Fernando Díaz
EUI: Escuela Universitaria de Informática, University of Valladolid, Plaza Santa Eulalia,
9-11, 40005, Segovia, Spain
e-mail: fdiaz@infor.uva.es

M.P. Rocha et al. (Eds.): IWPACBB 2010, AISC 74, pp. 9–16, 2010.
springerlink.com   © Springer-Verlag Berlin Heidelberg 2010

context, sample-based clustering is one of the most common methods for discovering disease subtypes as well as unknown taxonomies. By revealing hidden structures in microarray data, cluster analysis can potentially lead to more tailored therapies for patients as well as better diagnostic procedures.

From a practical point of view, existing sample-based clustering methods can be (*i*) directly applied to cluster samples using all the genes as features (i.e., classical techniques such as K-means, SOM, HC, etc.) or (*ii*) executed after a set of informative genes are identified. The problem with the first approach is the signal-to-noise ratio (smaller than 1:10), which is known to seriously reduce the accuracy of clustering results due to the existence of noise and outliers of the samples [1]. To overcome such difficulties, particular methods can be applied to identify informative genes and reduce gene dimensionality prior to clustering samples in order to detect their phenotypes. In this context, both supervised and unsupervised informative gene selection techniques have been developed.

While supervised informative gene selection techniques often yield high clustering accuracy rates, unsupervised informative gene selection methods are more complex because they assume no a priori phenotype information being assigned to any sample [2]. In such a situation, two general strategies have been adopted to address the lack of prior knowledge: (*i*) unsupervised gene selection, this aims to reduce the number of genes before clustering samples by using appropriate statistical models and (*ii*) interrelated clustering, that takes advantage of utilizing the relationship between the genes and samples to perform gene selection and sample clustering simultaneously in an iterative paradigm. Following the second strategy for unsupervised informative gene selection (interrelated clustering), Ben-Dor *et al.* [3] present an approach based on statistically scoring candidate partitions according to the overabundance of genes that separate the different classes. Xing and Karp [1] use a feature filtering procedure for ranking features according to their intrinsic discriminability and irredundancy to other relevant features. Their clustering algorithm is based on the concept of a normalized cut for grouping samples in new reference partition. Von Heydebreck *et al.* [4] and Tang *et al.* [5] propose algorithms for selecting sample partitions and corresponding gene sets by defining an indicator of partition quality and a search procedure to maximize this parameter. Varma and Simon [6] describe an algorithm for automatically detecting clusters of samples that are discernable only in a subset of genes.

In this contribution we are focused in the evaluation a novel simulated annealing-based algorithm for iterative class discovery. The rest of the paper is structured as follows: Section 2 sketches the proposed method and introduces the relevant aspects of the technique. Section 3 presents the experimental setup carried out and the results obtained from a publicly available microarray data set. Section 4 comprises a discussion about the obtained results by the proposed technique. Finally, Section 5 summarizes the main conclusions extracted from this work.

2   Overview of the Iterative Class Discovery Algorithm

In this article we propose a simulated annealing-based algorithm for iterative class discovery that uses a novel fuzzy logic method for informative gene selection. The interrelated clustering process carried out is based on an iterative approach where possible clusters are randomly constructed and evaluated by following a probabilistic search and an optimization schema.

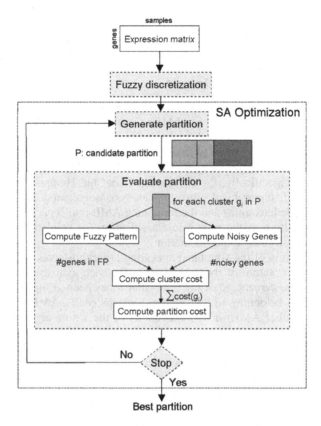

Fig. 1 Overview of the iterative class discovery method

Our clustering technique is not based on the distance between the microarrays belonging to each given cluster, but rather on the notion of *genetic coherence* of its own clusters. The genetic coherence of a given partition is calculated by taking into consideration the genes which share the same expression value through all the samples belonging to the cluster (which we term a *fuzzy pattern*), but discarding those genes present due to pure chance (herein referred to *noisy genes* of a fuzzy pattern). The proposed clustering technique combines both (*i*) the simplicity and good performance of a heuristic search method able to find good partitions in the space of all possible partitions of the set of samples with (*ii*) the robustness of fuzzy logic, able to cope with several levels of uncertainty and imprecision by using partial truth values. A global view of the proposed method is sketched in Figure 1. This figure shows how from the fuzzy discretization of the microarrays from raw dataset the method performs a stochastic search, looking for a "good partition" of microarrays in order to maximize the genetic coherence of each one cluster within the tentative partition.

3 Experimental Results

In this Section we evaluate the proposed algorithm on two public microarray datasets, herein referred to as HC-Salamanca dataset [7] and Armstrong dataset [8].

3.1 The HC-Salamanca Dataset

This dataset consists of bone marrow samples from 43 adult patients with de novo diagnosed acute myeloid leukemia (AML) – 10 acute promyelocytic leukemias (APL) with t(15;17), 4 AML with inv(16), 7 monocytic leukemias and 22 non-monocytic leukemias, according to the WHO classification. All samples contained more than 80% blast cells and they were analyzed using high-density oligonucleotide microarrays (specifically, the Affymetrix GeneChip Human Genome U133A Array) [7]. In [7], hierarchical clustering analysis segregated APL, AML with inv(16), monocytic leukemias and the remaining AML into separate groups, so we consider this partition as the reference classification for validating our proposed technique in the following experimentation.

As each execution of the simulated annealing algorithm gives a different result (due the stochastic nature of the search), then for each available microarray has been computed the percentage of the times that it has been grouped together with other microarrays belonging to the reference groups (APL, AML with inversion, Monocytic and Other AML) in ten executions of the whole algorithm. The percentage of times (on average) in which microarrays of each reference cluster have been grouped together with microarrays belonging to different classes is shown in each row of Table 1. This table can be interpreted as a confusion matrix numerically supporting the facts commented above, since the APL and Other-AML groups are the better identified pathologies (in an average percentage of 76.19% and 77.12% for all their samples and runs of the algorithm), followed by the monocytic leukemias (with an average percentage of 51.73%). As mentioned above, the group of AML with inversion is confused in a mean percentage of 33.66% and 32.06% with samples from monocytic and Other-AML groups, respectively. If we consider that the highest percentage for each microarray determines the cluster to which it belongs, the final clustering obtained by our simulated annealing-based algorithm is shown in Table 2.

Table 1 Confusion matrix for the HC-Salamanca dataset

		Predicted class			
		APL	*Inv*	*Mono*	*Other*
	APL	76.19%	2.71%	2.18%	18.92%
True class	*Inv*	7.79%	26.49%	33.66%	32.06%
	Mono	3.11%	17.81%	51.73%	27.35%
	Other	8.62%	5.56%	8.70%	77.12%

Table 2 Final clustering for the HC-Salamanca dataset

Cluster	Samples in cluster
APL	APL-05204, APL-10222, APL-12366, APL-13058, APL-13223, APL-14217, APL-14398, APL-16089, APL-16739, APL-17074, *Other-00139*
Mono	*Inv-00355*, *Inv-10891*, Mono-06667, Mono-09949, Mono-12361, Mono-13701, Mono-13774, Mono-13850, Mono-14043
Other	*Inv-00185*, *Inv-07644*, Other-00170, Other-06209, Other-07297, Other-09376, Other-09875, Other-10232, Other-10557, Other-11567, Other-12570, Other-13296, Other-13451, Other-14399, Other-14698, Other-14735, Other-15443, Other-15833, Other-16221, Other-16942, Other-16973, Other-17099, Other-17273

Assuming as "ground truth" the clustering given by authors in [7], the performance of the clustering process can be tested by comparing the results given in both tables. Some commonly used indices such as the *Rand index* and the *Jaccard* coefficient have been defined to measure the degree of similarity between two partitions. For the clustering given by our experiment, the *Rand index* was 0.90 and the *Jaccard coefficient* was 0.77.

3.2 The Armstrong Dataset

In [8] the authors proposed that lymphoblastic leukemias with MLL translocations (mixed-lineage leukemia) constitute a distinct disease, denoted as MLL, and show that the differences in gene expression are robust enough to classify leukemias correctly as MLL, acute lymphoblastic leukemia (ALL) or acute myeloid leukemia (AML). The public dataset of this work, herein referred to as the Armstrong dataset, has been also used to test our proposal. The complete group of samples consists of 24 patients with B-Precursor ALL (ALL), 20 patients with MLL rearranged B-precursor ALL (MLL) and 28 patients with acute myeloid leukemia (AML). All the samples were analyzed using the Affymetrix GeneChip U95a which contains 12600 known genes.

Table 3 Confusion matrix for the Armstrong dataset

		Predicted class		
		ALL	AML	MLL
True class	ALL	65.88%	5.16%	28.95%
	AML	4.42%	86.40%	9.18%
	MLL	34.74%	12.85%	52.41%

The percentage of times (on average) in which microarrays of each reference cluster have been grouped together with microarrays of different classes (across the ten executions of the algorithm) is shown in Table 3. These percentages can be considered as an estimation of the overlapping area of the membership functions of any two potential groups in the sector associated to a true class.

Table 4 Final clustering for the Armstrong dataset

Cluster	Samples in cluster
ALL	ALL-01, ALL-02, ALL-04, ALL-05, ALL-06, ALL-07, ALL-08, ALL-09, ALL-10, ALL-11, ALL-12, ALL-13, ALL-14, ALL-15, ALL-16, ALL-17, ALL-18, ALL-19, ALL-20, ALL-58, ALL-59, ALL-60, *MLL-25, MLL-32, MLL-34, MLL-62*
AML	*ALL-03*, AML-38, AML-39, AML-40, AML-41, AML-42, AML-43, AML-44, AML-46, AML-47, AML-48, AML-49, AML-50, AML-51, AML-52, AML-53, AML-54, AML-55, AML-56, AML-57, AML-65, AML-66, AML-67, AML-68, AML-69, AML-70, AML-71, AML-72
MLL	*ALL-61, AML-45*, MLL-21, MLL-22, MLL-23, MLL-24, MLL-26, MLL-27, MLL-28, MLL-29, MLL-30, MLL-31, MLL-33, MLL-35, MLL-36, MLL-37, MLL-63, MLL-64

As in the HC-Salamanca dataset, if the highest percentage for each sample determines the cluster of the microarray, the final clustering obtained by our simulated annealing-based algorithm is shown in Table 4. As in the previous experiment, assuming the clustering given by authors in [8] is the "ground truth", the *Rand index* and the Jaccard coefficient for experiments carried out are 0.89 and 0.72, respectively.

4 Discussion

The aim of the experiments reported in the previous section is to test the validity of the proposed clustering method. Dealing with unsupervised classification, it is very difficult to test the ability of a method to perform the clustering since there is no supervision of the process. In this sense, the classification into different groups proposed by the authors in [7, 8] is assumed to be the reference partition of samples in our work. This assumption may be questionable in some cases, since the reference groups are not well established. For example, in the HC-Salamanca dataset the AML with inversion group is established by observation of the karyotype of cancer cells, but there is no other evidence (biological, genetic) suggesting that this group corresponds to a distinct disease. Even so, the assumption of these prior partitions as reference groups is the only way to evaluate the similarity (or dissimilarity) of the results computed by the proposed method based on existing knowledge. As it turns out, there is no perfect match among the results of our proposed method and the reference partitions, but they are compatible with the current knowledge of each dataset. For example, for the HC-Salamanca dataset the better characterized groups are the APL and Other-AML groups, the worst is the AML with inversion group, and there is some confusion of the monocytic AML with the AML with inversion and Other-AML groups. These results are compatible with the state-of-the-art discussed in [7], where the APL group is the better characterized disease (it can be considered as a distinct class), the monocytic AML is a promising disease, the AML with inversion in chromosome 16 is the weaker class, and the Other-AML group acts as the dumping ground for the rest of samples which are not similar enough to the other possible classes. For the

Armstrong dataset, the AML group is clearly separated from the MLL and ALL groups. It is not surprising since the myeloid leukemia (AML) and lymphoblastic leukaemias (MLL and ALL) represent distinct diseases. Some confusion is present among ALL and MLL groups, but this result is compatible with the assumption (which the authors test in [8]) that the MLL group is a subtype of the ALL disease.

5 Conclusion

The simulated annealing-based algorithm presented in this work is a new algorithm for iterative class discovery that uses fuzzy logic for informative gene selection. An intrinsic advantage of the proposed method is that, assuming the percentage of times in which a given microarray has been grouped with samples of other potential classes, the degree of membership of that microarray to each potential group can be deduced. This fact allows a fuzzy clustering of the available microarrays which is more suitable for the current state-of-the-art in gene expression analysis, since it will be very unlikely to state (without uncertainty) that any available microarray only belongs to a unique potential cluster. In this case, the proposed method can help to assess the degree of affinity of each microarray with potential groups and to guide the analyst in the discovery of new diseases.

In addition, the proposed method is also an unsupervised technique for gene selection when it is used in conjunction with the concept of discriminant fuzzy pattern (DFP) introduced in [9]. Since the selected genes depend on the resulting clustering (they are the genes in the computed DFP obtained from all groups) and the clustering is obtained by maximizing the cost function (which is based on the notion of genetic coherence and assessed by the number of genes in the fuzzy pattern of each cluster), then the selected genes jointly depend on all the genes in the microarray, and the proposed method can be also considered a multivariate method for gene selection.

Finally, the proposed technique, in conjunction with our previous developed GENECBR platform [10], represents a more sophisticated tool which integrates three main tasks in expression analysis: clustering, gene selection and classification. In this context, all the proposed methods are non-parametric (they do not depend on assumptions about the underlying distribution of available data), unbiased with regard to the basic computational facility used to construct them (the notion of fuzzy pattern) and with the ability to manage imprecise (and hence, uncertain) information, which is implicit in available datasets in terms of degree of membership to linguistic labels (expressions levels, potential categories, etc.).

Acknowledgements

This work is supported in part by the project Development of computational tools for the classification and clustering of gene expression data in order to discover meaningful biological information in cancer diagnosis (ref. VA100A08) from JCyL (Spain).

References

1. Xing, E.P., Karp, R.M.: CLIFF: clustering of high-dimensional microarray data via iterative feature filtering using normalized cuts. Bioinformatics 17, S306–S315 (2001)
2. Jiang, D., Tang, C., Zhang, A.: Cluster analysis for gene expression data: a survey. IEEE T. Knowl. Data En. 16, 1370–1386 (2004)
3. Ben-Dor, A., Friedman, N., Yakhini, Z.: Class discovery in gene expression data. In: Proceedings of the Fifth Annual International Conference on Computational Biology. ACM, Montreal (2001)
4. von Heydebreck, A., Huber, W., Poustka, A., Vingron, M.: Identifying splits with clear separation: a new class discovery method for gene expression data. Bioinformatics 17, S107–S114 (2001)
5. Tang, C., Zhang, A., Ramanathan, M.: ESPD: a pattern detection model underlying gene expression profiles. Bioinformatics 20, 829–838 (2004)
6. Varma, S., Simon, R.: Iterative class discovery and feature selection using Minimal Spanning Trees. BMC Bioinformatics 5, 126 (2004)
7. Gutiérrez, N.C., López-Pérez, R., Hernández, J.M., Isidro, I., González, B., Delgado, M., Fermiñán, E., García, J.L., Vázquez, L., González, M., San Miguel, J.F.: Gene expression profile reveals deregulation of genes with relevant functions in the different subclasses of acute myeloid leukemia. Leukemia 19, 402–409 (2005)
8. Armstrong, S.A., Staunton, J.E., Silverman, L.B., Pieters, R., den Boer, M.L., Minden, M.D., Sallan, S.E., Lander, E.S., Golub, T.R., Korsmeyer, S.J.: MLL translocations specify a distinct gene expression profile that distinguishes a unique leukemia. Nat. Genet. 30, 41–47 (2002)
9. Díaz, F., Fdez-Riverola, F., Corchado, J.M.: geneCBR: a case-based reasoning tool for cancer diagnosis using microarray data sets. Comput. Intell. 22, 254–268 (2006)
10. Glez-Peña, D., Díaz, F., Hernández, J.M., Corchado, J.M., Fdez-Riverola, F.: geneCBR: a translational tool for multiple-microarray analysis and integrative information retrieval for aiding diagnosis in cancer research. BMC Bioinformatics 10, 187 (2009)

Automatic Workflow during the Reuse Phase of a CBP System Applied to Microarray Analysis

Juan F. De Paz, Ana B. Gil, and Emilio Corchado

Abstract. The application of information technology in the field of biomedicine has become increasingly important over the last several years. The different possibilities for the workflow in the microarray analysis can be huge and it would be very interesting to create an automatic process for establishing the workflows. This paper presents an intelligent dynamic architecture based on intelligent organizations for knowledge data discovery in biomedical databases. The multi-agent architecture incorporates agents that can perform automated planning and find optimal plans. The agents incorporate the CBP-BDI model for developing the automatic planning that makes possible to predict the efficiency of the workflow beforehand These agents propose a new reorganizational agent model in which complex processes are modelled as external services.

Keywords: Multiagent Systems, microarray, Case-based planning.

1 Introduction

The continuous growth of techniques for obtaining cancerous samples, specifically those using microarray technologies, provides a great amount of data. Microarray has become an essential tool in genomic research, making it possible to investigate global genes in all aspects of human disease [4]. Expression arrays [5] contain information about certain genes in a patient's samples. These data have a high dimensionality and require new powerful tools.

This paper presents an innovative solution to model reorganization systems in biomedical environments. It is based on a multi-agent architecture that can integrate Web services, and incorporates a novel planning mechanism that makes it possible to determine workflows based on existing plans and previous results. The

Juan F. De Paz · Ana B. Gil · Emilio Corchado
Departamento Informática y Automática
Universidad de Salamanca
Plaza de la Merced s/n, 37008, Salamanca, Spain
e-mail: {fcofds, abg, escorchado}@usal.es

M.P. Rocha et al. (Eds.): IWPACBB 2010, AISC 74, pp. 17–24, 2010.
springerlink.com   © Springer-Verlag Berlin Heidelberg 2010

core of the system presented in this paper is a CBP-BDI (Case-based planning) (Belief Desire Intention) agent [3] specifically designed to act as Web services co-ordinator, making it possible to reduce the computational load for the agents in the organization and expedite the classification process. CBP-BDI agents [2] make it possible to formalize systems by using a new planning mechanism that incorpo-rates graph theory as a reasoning engine to generate plans. The system was specifically applied to case studies consisting of the classification of cancers from microarrays. The multi-agent system developed incorporates novel strategies for data analysis and microarray data classification.

The next section describes the expression analysis problem. Section 2 presents a case study consisting of a distributed multi-agent system for cancer detection scenarios. Finally section 3 presents the results and conclusions obtained.

2 Self-Adaptive Multiagent System for Expression Analysis

Nowadays, it is essential to have software solutions that enforce autonomy, ro-bustness, flexibility and adaptability of the system to develop. The dynamic agent organizations that auto-adjust themselves to obtain advantages from their envi-ronment seem to be a technology that is more than suitable for coping with the de-velopment of this type of system. The integration of multi-agent systems with SOA (Service Oriented Architecture) and Web Services approaches has been re-cently explored [7]. Some developments are centered on communication between these models, while others are centered on the integration of distributed services, especially Web Services, into the structure of the agents. [8] Oliva et al. [8] have developed a java-based framework to create SOA and Web Services compliant applications, which are modelled as agents.

The approach presented in this paper is an organizational model for biomedical environments based on a multi-agent dynamic architecture that incorporates agents capable of generating plans for analyzing large amounts of data. The core of the system is a novel mechanism for the implementation of the stages of CBP-BDI mechanisms through Web services that provides a dynamic self-adaptive behaviour to reorganize the environment. The types of agents are distrib-uted in layers within the system according to their functionalities. The agent layers constitute the core and define a virtual organization for massive data analysis:

- Organization: The agents will be responsible for conducting the analysis of information following the CBP-BDI [2] reasoning model. The agents from the organizational layer should be initially configured for the differ-ent types of analysis that will be performed, given that these analyses vary according to the available information and the search results.
- Analysis: The agents in the analysis layer are responsible for selecting the configuration and the flow of services best suited for the problem at hand. They communicate with Web services to generate results. The agents of this layer follow the CBP-BDI [2] reasoning model. The work-flow and configuration of the services to be used is selected with graphs, using information that corresponds to the previously executed plans.

- The Controller agent manages the agents available in the different layers of the multiagent system. It allows the registration of agents in the layers, as well as their use in the organization.
- Analysis Services: The analysis services are services used by analysis agents for carrying out different tasks. The analysis services include services for pre-processing, filtering, clustering and extraction of knowledge.

2.1 Coordinator CBP-BDI Agent

The agents in the analysis layer have the capacity to learn from the analysis carried out in previous procedures. They adopt the CBP reasoning model, a specialization of case-based reasoning (CBR) [1]. CBR systems solve new problems by adapting solutions that have been used to solve similar problems in the past, and learning from each new experience. A CBR manages cases (past experiences) to solve new problems. The way cases are managed is known as the CBR cycle, and consists of four sequential phases: retrieve, reuse, revise and retain. CBP is the idea of planning as remembering [2]. In CBP, the solution proposed to solve a given problem is a plan, so this solution is generated by taking into account the plans applied to solve similar problems in the past [6]. The CBP-BDI agents stem from the BDI model [9] and establish a correspondence between the elements from the BDI model and the CBP systems. Fusing the CBP agents together with the BDI model and generating CBP-BDI agents makes it possible to formalize the available information, the definition of the goals and actions that are available for resolving the problem, and the procedure for resolving new problems by adopting the CBP reasoning cycle. Agent plan is the name we give to a sequence of actions that, from a current state e_0, defines the path of states through which the agent passes in order to reach the other world state.

$$p_n(e_0) = e_n = a_n(e_{n-1}) = \cdots = (a_n \circ \cdots \circ a_1)(e_0) \quad p_n \equiv a_n \circ \cdots \circ a_1 \tag{1}$$

Based on this representation, the CBP-BDI coordinator agents combine the initial state of a case, the final state of a case with the goals of the agent, and the intentions with the actions that can be carried out in order to create plans that make it possible to reach the final state. The actions that need to be carried out are services, making a plan an ordered sequence of services. It is necessary to facilitate the inclusion of new services and the discovery of new plans based on existing plans. Services correspond to the actions that can be carried out and that determine the changes in the initial problem data. The plan actions correspond to services and the order in which the actions are applied correspond to the order for executing services. As such, an organizational plan is defined by the services that comprise it and by the order of applying each of those services.

The information corresponding to each plan is represented in bidimensional arrays as shown in the chart in figure 1. The chart lists the plans in rows while the colums represent the links between the services that comprise a plan so that S_{ij} represents the execution of service j occurring after service call i. The second row shows the plan comprised of services a_2, a_1, with an initial connection S_{02} that executes service a_2 at the initial stage. The columns for service S_{2x} provide the connection with the subsequent service, i.e., S_{21}, for which service a_1 is executed. Lastly, column S_{1x} executes action S_{1f}.

Actions/Services

		S_{01}	S_{02}	...	S_{12}	S_{13}	...	S_{1f}	...	S_{21}	S_{23}	...	S_{2f}	...	S_{i1}	S_{ij}	...	S_{if}		
Plans	a_1	v		...			...	v	...			...		...			...	v	v_1	Efficiency
	a_2a_1		v	...			...	v	...	v		...		...			...		v_2	
	a_1a_2	v		...		v		...			...	v		...			...		v_3	

Fig. 1 Plans and plan actions carried out through a concatenation of services

Based on the information corresponding to previous experiences (plans already executed) a new plan is generated. To do so, the cases with the greatest and least efficiency with regards to the current problem are retrieved, and the CBP reasoning cycle is initiated according to the BDI specifications. This way, each plan is represented by the following expression:

$$p = \{(a_f \circ a_k \circ \cdots \circ a_i \circ a_0)(e_0) = (S_{kf} \circ \cdots \circ S_{0i})(e_0)\} \tag{2}$$

where e_0 represents the initial state that corresponds to the initial value of each probe. As each of the selected services are executed, different states e_i, are reached, which contain the new set of probes produced by the application of services.

2.1.1 Retrieve

During the retrieval stage, the plans with the greatest and least efficiency are selected from among those that have been applied. Microarrays are composed of probes that represent variables that mark the level of significance of specific genes. The retrieval of those cases is performed in one of two ways according to the case study. To retrieve cases, it is important to consider whether there has been a previous analysis of a case study with similar characteristics. If so, the corresponding plans for the same case study are selected.

If there are no plans that correspond to the same case study, or if the number of plans is insufficient, the plans corresponding ot the most similar case study are retrieved. The selection of the most similar case study is performed according to the cosine distance applied to the following set of variables: Number of probes, Number of cases, Coefficient of the Pearson variation [12] for e_0.

The number of efficient and inefficient cases selected is predetermined so that at the end of this stage the following set of elements is obtained:

$$P = \{P_e\{p_1^e, ..., p_n^e\} \cup P_i\{p_1^i, ..., p_n^i\}\} \tag{3}$$

P_e represents the set of efficient plans and P_i represents the set of inefficient plans. Once the plans have been retrieved, a new efficient plan is generated in the next phase.

2.1.2 Reuse

This phase takes the plans P obtained in the retrieval phase and generates a new, more efficient plan. The new plan is built according to the efficiency of the actions as estimated by the overall efficiency vi of the plan. Estimating the efficiency of each action is done according to the model defined by the decision trees for selecting significant nodes [3]. This way, estimating the efficiency of each action is carried out according to the expression (3). This expression is referred to as the winning rate and depends on both node S and the selected attribute B.

$$G(S,B) = I(S) - \sum_{i=1}^{t} \frac{|S_i|}{|S|} I(S_i) \tag{4}$$

where S represents a node that, in this case, will always be the root node of the tree, B is the condition for the existing action, S_i represents child node i from node S, $|S_i|$ the number of cases associated with the child node S_i. The function $I(S)$ represents gain and is defined as follows

$$I(S) = -\sum_{j=1}^{n} f_j^S \cdot \log(f_j^S) \tag{5}$$

where f_j^S represents the frequency relative to class C_j in S, $f_j^S = \frac{n_j^S}{N^S}$, n_j^S the number of elements from class C_j in S and N^S the total number of elements. In this case, Cj={efficient, inefficient}.

The gain ratio G determines the level of importance for each action by distinguishing between an efficient and an inefficient plan. High values for the gain ratio indicate that the action should be included in a plan if it involves an action to be carried out in an efficient plan, otherwise it should be eliminated.

A new table listing gain data is formed according to the values of the gain ratio and the efficiency associated with each plan. A new flow of execution for each action is created from the gains table. The gains uses the following formula to establish a value for the significance of each of the actions carried out in each plan:

$$T(S_{ij}, k) = G'(S, S_{ij}) \cdot \bar{v}_k \tag{6}$$

where G′ contains the values of G that are normalized between 0 and 1 with the values being inverted (the maximum value corresponds to 0 and the minimum to 1) and $\bar{v}$ contains the average value of efficiency for the plans with a connection ij. Each connection ij presents an influence in the final efficiency of the plan that is represented as t_{ijk} .

Once the graph for the plans has been constructed, the minimal route that goes from the start node to the end node is calculated. In order to calculate the shortest/longest route, the Dijkstra algorithm is applied since there are implementations for the order n*log n. To apply this algorithm, it is necessary to add to each of the edges the absolute value of the edge with a higher negative absolute value, in order to remove from the graph those edges with negative values.

2.1.3 Revise and Retain

The revise phase is carried out automatically according to the final efficiency obtained. The different analyses are associated with different measures of efficiency that measure the final quality of the results obtained, making it unnecessary to perform a manual revision. During the retain phase, the plan is stored in the memory of plans. If a plan with the same flow of execution of services for the same case study already exists, only the information from the plan with the highest quality will be stored. This makes it possible to limit the size of the case memory and select only those plans with certain parameters such as level of significance, correlation coefficient, percentile, etc.

3 Results and Conclusions

This paper has presented a self-adaptive organization based on a multiagent architecture and its application to a real problem. In 2006, there were approximately 64,122 men and women alive in the United States who had a history of cancer of the stomach: 36,725 men and 27,397 women. The age-adjusted incidence rate was 7.9 per 100,000 men and women per year while the fatality rate was 4.0 per 100,000 [11]. The data for gastric cancer were obtained with a HG U133 plus 2.0 chip and corresponded to 234 patients affected by this cancer in 3 different parts of the organism (primary gastric cancer, skin and others) [10]. The data were obtained from a public repository at http://www.ebi.ac.uk.

The experiment consisted of evaluating the services distribution system in the filtering agent for the case study that classified patients affected by different types of cancer. According to the identification of the problem described in table 1, the filtering agent selected the plans with the greatest efficiency, considering the different execution workflows for the services that are in the plans.

The filtering agent in the analysis layer selects the configuration parameters between a specific set of pre-determined values, when it has been told to explore the parameters. Otherwise, for a specific plan, it selects the values that have provided better results based on the measure of the previously established efficiency. The different configurations used are listed in table 1. A depth analysis of these techniques can be found in our previous work [10]. The last columns of the table list the final efficiency obtained based a measure and the type of plan (efficient or inefficient). A value of 1 in the Class column indicates that the plan is efficient while a 0 indicates that the plan is inefficient. The remaining value indicates the order of execution of the services.

Table 1 Efficiency of the plans

Plan	Variability (z)	Uniform (α)	Correlation (α)	Cutoff	Efficiency	Class
p_1	1	2	3		0.14641098	1
p_2	1	2	3	4	1	0
p_3	1		2		0.24248635	1
p_4		1	2		0.14935538	1
p_5	3	1	2		0.15907924	1
p_6	1				0.96457118	0
p_7				1	1	0
p_8			1	2	1	0

The diagnostic agent at the Organization level is in charge of selecting which filtering agent to use. The filtering agent, in turn, automatically and in accordance with the efficiency of each of the plans listed in table 1 and the equation (5), selects the flow of services that best adapts to the case study. In accordance with the equation defined in (5) and the information regarding efficiency provided by table 1, the relevance for the execution of each of the services is calculated.

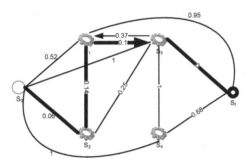

Fig. 2 Decision tree for classifying patients

Figure 2 displays the directed graph that was obtained. The final path that is followed is shown in bold. The new estimated plan is comprised of the sequence of actions S_{02}, S_{21}, S_{13}, S_{3f}.

It is clear that the path followed in the plan that was obtained does not coincide with any path previously applied, although the services that it contains presents an efficiency similar to that given by plan p_1 as shown in table 1. The efficiency obtained in this execution is 0.14458.

The system presented in this study provides a novel mechanism for a global coordination in highly changing environments. The mechanism is capable of automatic reorganization and is especially useful for decision making in systems that use agreement technologies. The system was applied to a case study in a biomedical environment and can be easily extended to other environments with similar characteristics.

Acknowledgments. This development has been partially supported by the projects JCyL SA071A08, of the Junta of Castilla and León (JCyL): [BU006A08], the project of the Spanish Ministry of Education and Innovation [CIT-020000-2008-2] and [CIT-020000-2009-12], and Grupo Antolin Ingenieria, S.A., within the framework of project MAGNO2008 - 1028.- CENIT also funded by the same Government Ministry.

References

[1] Kolodner, J.: Case-Based Reasoning. Morgan Kaufmann, San Francisco (1993)

[2] Glez-Bedia, M., Corchado, J.: A planning strategy based on variational calculus for deliberative agents. Computing and Information Systems Journal 10(1), 2–14 (2002)

[3] Kohavi, R., Ross Quinlan, R.: Decision Tree Discovery Handbook of Data Mining and Knowledge Discovery, pp. 267–276. Oxford University Press, Oxford (2002)

[4] Quackenbush, J.: Computational analysis of microarray data. Nature Review Genetics 2(6), 418–427 (2001)

[5] Affymetrix,
http://www.affymetrix.com/support/technical/datasheets/hgu133arrays_datasheet.pdf

[6] Corchado, J.M., Bajo, J., De Paz, Y., Tapia, D.I.: Intelligent Environment for Monitoring Alzheimer Patients, Agent Technology for Health Care. Decision Support Systems 44(2), 382–396 (2008)

[7] Ardissono, L., Petrone, G., Segnan, M.: A conversational approach to the interaction with Web Services. Computational Intelligence, vol. 20, pp. 693–709. Blackwell Publishing, Malden (2004)

[8] Oliva, E., Natali, A., Ricci, A., Viroli, M.: An Adaptation Logic Framework for {J}ava-based Component Systems. Journal of Universal Computer Science 14(13), 2158–2181 (2008)

[9] Bratman, M.: Intention, Plans and Practical Reason. Harvard U.P., Cambridge (1987)

[10] Corchado, J.M., De Paz, J.F., Rogríguez, S., Bajo, J.: Model of experts for decision support in the diagnosis of leukemia patients. Artificial Intelligence in Medicine 46(3), 179–200 (2009)

[11] Horner, M.J., Ries, L.A.G., Krapcho, M., Neyman, N., Aminou, R., Howlader, N., Altekruse, S.F., Feuer, E.J., Huang, L., Mariotto, A., Miller, B.A., Lewis, D.R., Eisner, M.P., Stinchcomb, D.G., Edwards, B.K. (eds.): SEER Cancer Statistics Review, 1975-2006, National Cancer Institute (2009),
http://seer.cancer.gov/csr/1975_2006/

[12] Kuo, C.D., Chen, G.Y., Wang, Y.Y., Hung, M.J., Yang, J.L.: Characterization and quantification of the return map of RR intervals by Pearson coefficient in patients with acute myocardial infarction. Autonomic Neuroscience 105(2), 145–152 (2003)

A Comparative Study of Microarray Data Classification Methods Based on Ensemble Biological Relevant Gene Sets

Miguel Reboiro-Jato, Daniel Glez-Peña, Juan Francisco Gálvez,
Rosalía Laza Fidalgo, Fernando Díaz, and Florentino Fdez-Riverola

Abstract. In this work we study the utilization of several ensemble alternatives for the task of classifying microarray data by using prior knowledge known to be biologically relevant to the target disease. The purpose of the work is to obtain an accurate ensemble classification model able to outperform baseline classifiers by introducing diversity in the form of different gene sets. The proposed model takes advantage of WhichGenes, a powerful gene set building tool that allows the automatic extraction of lists of genes from multiple sparse data sources. Preliminary results using different datasets and several gene sets show that the proposal is able to outperform basic classifiers by using existing prior knowledge.

Keywords: microarray data classification, ensemble classifiers, gene sets, prior knowledge.

1 Introduction and Motivation

The advent of microarray technology has become a fundamental tool in genomic research, making it possible to investigate global gene expression in all aspects of human disease. In particular, cancer genetics based on the analysis of cancer genotypes, provides a valuable alternative to cancer diagnosis in both theory and practice [1]. In this context, the automatic classification of cancer patients has been a

Miguel Reboiro-Jato · Daniel Glez-Peña · Juan Francisco Gálvez · Rosalía Laza Fidalgo ·
Florentino Fdez-Riverola
ESEI: Escuela Superior de Ingeniería Informática, University of Vigo,
Edificio Politécnico, Campus Universitario As Lagoas s/n, 32004, Ourense, Spain
e-mail: {mrjato, dgpena, galvez, rlaza, riverola}@uvigo.es

Fernando Díaz
EUI: Escuela Universitaria de Informática, University of Valladolid, Plaza Santa Eulalia,
9-11, 40005, Segovia, Spain
e-mail: fdiaz@infor.uva.es

M.P. Rocha et al. (Eds.): IWPACBB 2010, AISC 74, pp. 25–32, 2010.
springerlink.com   © Springer-Verlag Berlin Heidelberg 2010

promising approach in cancer diagnosis since the early detection and treatment can substantially improve the survival rates. For this task, several computational methods (statistical and machine learning) have been proposed in the literature including linear discriminant analysis (LDA), Naïve-Bayes classifier (NBC), learning vector quantization (LVQ), radial basis function (RBF) networks, decision trees, probabilistic neural networks (PNNs) and support vector machines (SVMs) among others [2]. In the same line, but following the assumption that a classifier ensemble system is more robust than an excellent single classifier [3], some researchers have also successfully applied different classifier ensemble systems to deal with the classification of microarray datasets [4].

In addition to predictive performance, there is also hope that microarray studies uncover molecular disease mechanisms. However, in many cases the molecular signatures discovered by the algorithms are unfocused form a biological point of view [5]. In fact, they often look more like random gene lists than biologically plausible and understandable signatures. Another shortcoming of standard classification algorithms is that they treat gene-expression levels as anonymous attributes. However, a lot is known about the function and the role of many genes in certain biological processes.

Although numerical analysis of microarray data is considerable consolidated, the true integration of numerical analysis and biological knowledge is still a long way off [6]. The inclusion of additional knowledge sources in the classification process can prevent the discovery of the obvious, complement a data-inferred hypothesis with references to already proposed relations, help analysis to avoid overconfident predictions and allow us to systematically relate the analysis findings to present knowledge [7]. In this work we would like to incorporate relevant gene sets obtained from WhichGenes [8] in order to make predictions easy to interpret in concert with incorporated knowledge. The study carried out aims to borrow information from existing biological knowledge to improve both predictive accuracy and interpretability of the resulting classifiers.

The rest of the paper is structured as follows: Section 2 presents a brief review about the use of ensemble methods for classifying microarray data. Section 3 describes the selected datasets and base classifiers for the current study, together with the choice of gene sets and the different approaches used for ensemble creation. Finally Section 4 discusses the reported results and concludes the paper.

2   Related Work

Although much research has been performed on applying machine learning techniques for microarray data classification during the past years, it has been shown that conventional machine learning techniques have intrinsic drawbacks in achieving accurate and robust classifications. In order to obtain more robust microarray data classification techniques, several authors have investigated the benefits of this approach applied to genomic research.

Díaz-Uriarte and Alvarez de Andrés [9] investigated the use of random forest for multi-class classification of microarray data and proposed a new method of gene selection in classification problems based on random forest. Using simulated and real microarray datasets the authors showed that random forest can obtain comparable performance to other methods, including DLDA, KNN, and SVM.

Peng [10] presented a novel ensemble approach based on seeking an optimal and robust combination of multiple classifiers. The proposed algorithm begins with the generation of a pool of candidate base classifiers based on the gene sub-sampling and then, it performs the selection of a sub-set of appropriate base classifiers to construct the classification committee based on classifier clustering. Experimental results demonstrated that the proposed approach outperforms both baseline classifiers and those generated by bagging and boosting.

Liu and Huang [11] applied Rotation Forest to microarray data classification using principal component analysis, non-parametric discriminant analysis and random projections to perform feature transformation in the original rotation forest. In all the experiments, the authors reported that the proposed approach outperformed bagging and boosting alternatives.

More recently, Liu and Xu [12] proposed a genetic programming approach to analyze multiclass microarray datasets where each individual consists of a set of small-scale ensembles containing several trees. In order to guarantee high diversity in the individuals a greedy algorithm is applied. Their proposal was tested using five datasets showing that the proposed method effectively implements the feature selection and classification tasks.

As a particular case in the use of ensemble systems, ensemble feature selection represents an efficient method proposed in [13] which can also achieve high classification accuracy by combining base classifiers built with different feature subsets. In this context, the works of [14] and [15] study the use of different genetic algorithms alternatives for performing feature selection with the aim of making classifiers of the ensemble disagree on difficult cases. Reported results on both cases showed improvements when compared against other alternatives.

Related with previous work, the aim of this study is to validate the superiority of different classifier ensemble approaches when using prior knowledge in the form of biological relevant gene sets. The objective is to improve the predictive performance of baseline classifiers.

3 Comparative Study

In order to carry out the comparative study, we apply several ensemble alternatives to classify three DNA microarray datasets involving various tumour tissue samples. With the goal of validate the study, we analyze the performance of different baseline classifiers and test our hypothesis using two different sources of information.

3.1 Datasets and Base Classifiers

We carry out the experimentation using three public leukemia datasets taken from the previous studies of Gutiérrez *et al* [16], Bullinger *et al* [17] and Valk *et al* [18]. We have selected samples from each dataset belonging to 4 different groups of acute myeloid leukemias including (*i*) promyelocytic (APL), (*ii*) inversion 16, (*iii*) monocytic and (*iv*) other AMLs. The distribution of samples is showed in Table 1.

Table 1 Distribution of microarray data samples belonging to the public datasets analyzed

	APL	Inv(16)	Monocytic	Other
Gutiérrez *et al*	10	4	7	22
Bullinger *et al*	19	14	64	177
Valk *et al*	7	10	7	51

In order to compare the performance obtained by the different ensemble approaches, we have selected four well-known classification algorithms: (*i*) Naïve Bayes (NB) learner is perhaps the most widely used method. Although its independence assumption is over-simplistic, studies have found NB to be very effective in a wide range of problems; (*ii*) IB3 represents a variant of the well-known nearest neighbour algorithms implementing a simple version of a lazy learner classifier; (*iii*) Support Vector Machines (SVMs) constitute a famous family of algorithms used for classification and regression purposes. Their mayor advantage is that their learning capacity does not degrade even if many characteristics exist, being especially applicable to microarray data; (*iv*) Random Forest (RFs) is a basic ensemble classifier that consists of many decision trees. The method combines bagging idea and random selection of features in order to construct a collection of decision trees with controlled variation.

3.2 Biological Knowledge Gene Sets

For the prior selection of gene sets that represent explicit information available the following sources of information have been used: (*i*) 33 metabolic sub-pathways related to existing cancers in SABiosciences (http://www.sabiosciences.com) previously analyzed in studies by [19] and [20] plus 4 groups extracted from the OMIM (*Online Mendelian Inheritance in Man*) database (http://www.ncbi.nlm.nih.gov/sites/entrez?db=omim) that correspond to various types of leukemia (myeloid, monocytoid, promyelocytic and general leukemia) and (*ii*) those pathways from KEGG (*Kyoto Encyclopedia of Genes and Genomes*) database grouped in both 'environmental information processing' and 'genetic information processing' categories.

3.3 Ensemble Alternatives

According to Kuncheva [3], several ensembles can be built by introducing variations at four different levels: (*i*) data level, (*ii*) feature level, (*iii*) classifier level and (*iv*) combination level.

First of all, by using different data subsets at data level or different feature subsets at feature level, the space of the problem can be divided into several areas where base classifiers can be trained. This divide-and-conquer strategy can simplify the problem, leading to improved performance of the base classifiers. Secondly, at classifier level, different types of classifiers can be used in order to take advantage of the strong points of each classifier type. Although many ensemble paradigms employ the same classification model, there is no evidence that one strategy is better than the other [3]. Finally, combination level groups the different ways of combining the classifier decisions.

In this study, base classifiers are trained with all the samples in each data set, so no work is performed at data level. The feature level is carried out by incorporating gene set data to the ensemble models. Each pathway or group of genes is used as a feature selection, so microarray data will be filtered to keep only the expression level of those genes belonging to some group before training base classifiers.

In order to construct the final ensemble, our approach consists on two sequential steps: (*i*) *classifier selection*, in which each simple classifier is initially trained with each gene set following a stratified 10-fold cross-validation process for estimating its performance and (*ii*) *classifier training*, where the selected pairs of simple_classifier/gene_set are trained with the whole data set. All the different strategies proposed in this study for the selection of promising classifiers are based on the value of the kappa statistic obtained for each simple_classifier/gene_set pair in the first step. The proposed heuristics are the following:

- *All classifiers* [AC]: every simple_classifier/gene_set pair is used for constructing the final ensemble.
- *All gene sets* [AG]: for each gene set, the simple_classifier/gene_set pair with best kappa value is selected for constructing the final ensemble.
- *Best classifiers without type* [BCw/oT_%]: a global threshold is calculated as a percentage of the best kappa value obtained by the winner simple_classifier/gene_set pair. Those pairs with a kappa value equal or higher than the computed threshold are selected.
- *Best classifier by type* [BCbyT_%]: as in the previous heuristic a given threshold is calculated, but in this case there is a threshold for each simple classifier type.

The form in which the final output of the ensemble is calculated is also based on the kappa statistic. The combination approach used on for the proposed ensembles is a weighted majority vote where the weight of each vote is the corresponding classifier's kappa value.

4 Experimental Results and Discussion

In order to evaluate the heuristics defined in the previous section, a comparative study was carried out using two different sources of information (OMIM and KEGG) in order to classify 392 samples belonging to four classes coming from three real data scts. In addition, the four simple base classifiers used for the ensemble generation (IB3, NBS, RF, SVM) where also tested individually, using as feature selection both those genes included in the OMIM gene sets plus those genes being part of the KEGG gene sets. Classification tests were performed using a stratified 10-fold cross-validation. Tables 2 and 3 summarize the results obtained from the experimentation carried out showing only those classifiers with better performance.

Table 2 presents the accuracy and kappa values achieved by each classifier using KEGG gene sets as prior knowledge. As it can be observed, BCbyT heuristic generally exhibits good performance regardless of the data set. Additionally, BCw/oT heuristic also showed good performance, although in the Gutiérrez data set two single classifiers (IB3 and NBS) performed better than ensembles using this strategy.

Table 2 Classification result using KEGG gene sets

Classifier	Gutiérrez		Bullinger		Valk	
	Accuracy	Kappa	Accuracy	Kappa	Accuracy	Kappa
AC	76,74%	0,588	76,00%	0,292	76,28%	0,503
AG	79,07%	0,634	76,00%	0,299	75,18%	0,533
BCbyT_90%	81,40%	0,724	82,67%	0,373	77,01%	0,540
BCbyT_95%	83,72%	0,724	82,67%	0,329	**78,83%**	**0,574**
BCw/oT_60%	79,07%	0,635	**80,00%**	**0,476**	77,37%	0,556
BCw/oT_75%	79,07%	0,635	81,33%	0,367	76,64%	0,555
BCw/oT_85%	76,74%	0,612	80,00%	0,403	75,55%	0,546
IB3	**83,72%**	**0,756**	69,33%	0,369	67,52%	0,410
NBS	81,40%	0,679	73,33%	0,269	74,09%	0,530
RF	72,09%	0,533	68,00%	0,123	69,71%	0,337
SVM	51,16%	0,000	68,00%	0,000	64,60%	0,000

Table 3 Classification result using OMIM gene sets

Classifier	Gutiérrez		Bullinger		Valk	
	Accuracy	Kappa	Accuracy	Kappa	Accuracy	Kappa
AC	76,74%	0,588	76,00%	0,343	74,82%	0,439
AG	76,74%	0,588	76,00%	0,343	76,28%	0,506
BCbyT_90%	81,40%	0,680	**82,67%**	**0,569**	75,91%	0,528
BCbyT_95%	86,05%	0,774	**82,67%**	**0,569**	75,91%	0,513
BCw/oT_60%	81,40%	0,680	80,00%	0,483	75,91%	0,521
BCw/oT_75%	**88,37%**	**0,809**	81,33%	0,526	75,91%	0,530
BCw/oT_85%	79,07%	0,672	80,00%	0,482	76,28%	0,555
IB3	76,74%	0,643	73,33%	0,451	67,52%	0,391
NBS	79,07%	0,634	76,00%	0,370	72,99%	0,510
RF	79,07%	0,658	74,67%	0,372	74,09%	0,420
SVM	51,16%	0,000	68,00%	0,000	77,74%	0,539

Table 3 presents the same experimentation but using the OMIM gene sets. Once again, BCbyT heuristic achieved good performance. Comparing its behaviour against single classifiers, performance of ensembles is even better than in the previous experimentation (using KEGG gene sets). BCw/oT heuristic also performs better with the OMIM gene set, being slightly superior to BCbyT heuristic. Ensembles using this strategy not only performed better than single classifiers, but also achieved the best kappa value in two of the three analyzed data sets.

To sum up, we can conclude that BCbyT heuristic performed as the best base classifier selection strategy, followed closely by *BCw/oT* heuristic. This fact backs up the following ideas: (*i*) depending on the data set there is not a single classifier able to achieve good performance in concert with the supplied knowledge and (*ii*) the presence of each classifier type in the final ensemble may improve the classification performance.

Regardless of the data set both *BCw/oT* and BCbyT heuristics behave uniformly performing better than single baseline classifiers. This circumstance confirms the fact that ensembles generally perform better than single classifiers, in this case, by taking advantage of using prior structured knowledge.

Acknowledgements. This work is supported in part by the project *MEDICAL-BENCH: Platform for the development and integration of knowledge-based data mining techniques and their application to the clinical domain* (TIN2009-14057-C03-02) from Ministerio de Ciencia e Innovación (Spain). D. Glez-Peña acknowledges Xunta de Galicia (Spain) for the program Ángeles Álvariño.

References

1. Golub, T.R., Slonim, D.K., Tamayo, P., Huard, C., Gaasenbeek, M., Mesirov, J.P., Coller, H., Loh, M.L., Downing, J.R., Caligiuri, M.A., Bloomfield, C.D., Lander, E.S.: Molecular classification of cancer: class discovery and class prediction by gene expression monitoring. Science 286, 531–537 (1999)
2. Ressom, H.W., Varghese, R.S., Zhang, Z., Xuan, J., Clarke, R.: Classification algorithms for phenotype prediction in genomics and proteomics. Frontiers in Bioscience 13, 691–708 (2008)
3. Kuncheva, L.I.: Combining Pattern Classifiers: Methods and Algorithms. Wiley Interscience, Hoboken (2004)
4. Liu, K.H., Li, B., Wu, Q.Q., Zhang, J., Du, J.X., Liu, G.Y.: Microarray data classification based on ensemble independent component selection. Computers in Biology and Medicine 39(11), 953–960 (2009)
5. Lottaz, C., Spang, R.: Molecular decomposition of complex clinical phenotypes using biologically structured analysis of microarray data. Bioinformatics 21(9), 1971–1978 (2005)
6. Cordero, F., Botta, M., Calogero, R.A.: Microarray data analysis and mining approaches. Briefings in Functional Genomics and Proteomics 6(4), 265–281 (2007)
7. Bellazzi, R., Zupan, B.: Methodological Review: Towards knowledge-based gene expression data mining. Journal of Biomedical Informatics 40(6), 787–802 (2007)
8. Glez-Peña, D., Gómez-López, G., Pisano, D.G., Fdez-Riverola, F.: WhichGenes: a web-based tool for gathering, building, storing and exporting gene sets with application in gene set enrichment analysis. Nucleic Acids Research 37(Web Server issue), W329–W334 (2009)
9. Díaz-Uriarte, R., Alvarez de Andrés, S.: Gene selection and classification of microarray data using random forest. BMC Bioinformatics 7, 3 (2006)
10. Peng, Y.: A novel ensemble machine learning for robust microarray data classification. Computers in Biology and Medicine 36(6), 553–573 (2006)
11. Liu, K.H., Huang, D.S.: Cancer classification using Rotation Forest. Computers in Biology and Medicine 38(5), 601–610 (2008)
12. Liu, K.H., Xu, C.G.: A genetic programming-based approach to the classification of multiclass microarray datasets. Bioinformatics 25(3), 331–337 (2009)
13. Opitz, D.: Feature selection for ensembles. In: Proceedings of 16th National Conference on Artificial Intelligence, Orlando, Florida (1999)
14. Kuncheva, L.I., Jain, L.C.: Designing classifier fusion systems by genetic algorithms. IEEE Transactions on Evolutionary Computation 4(4), 327–336 (2000)

15. Oliveira, L.S., Morita, M., Sabourin, R.: Feature selection for ensembles using the multi-objective optimization approach. Studies in Computational Intelligence 16, 49–74 (2006)
16. Gutiérrez, N.C., López-Pérez, R., Hernández, J.M., Isidro, I., González, B., Delgado, M., Fermiñán, E., García, J.L., Vázquez, L., González, M., San Miguel, J.F.: Gene expression profile reveals deregulation of genes with relevant functionsin the different subclasses of acute myeloid leukemia. Leukemia 19(3), 402–409 (2005)
17. Bullinger, L., Döhner, K., Bair, E., Fröhling, S., Schlenk, R.F., Tibshirani, R., Döhner, H., Pollack, J.R.: Use of gene-expression profiling to identify prognostic subclasses in adult acute myeloid leukemia. The New England Journal of Medicine 350(16), 1506–1516 (2004)
18. Valk, P.J., Verhaak, R.G., Beijen, M.A., Erpelinck, C.A., Barjesteh van Waalwijk van Doorn-Khosrovani, S., Boer, J., Beverloo, H., Moorhouse, M., van der Spek, P., Löwenberg, B., Delwel, R.: Prognostically useful gene-expression profiles in Acute Myeloid Leukemia. The New England Journal of Medicine 350(16), 1617–1628 (2004)
19. Tai, F., Pan, W.: Incorporating prior knowledge of predictors into penalized classifiers with multiple penalty terms. Bioinformatics 23(14), 1775–1782 (2007)
20. Wei, Z., Li, H.: Nonparametric pathway-based regression models for analysis of genomic data. Biostatistics 8(2), 265–284 (2007)

Predicting the Start of Protein α-Helices Using Machine Learning Algorithms

Rui Camacho, Rita Ferreira, Natacha Rosa, Vânia Guimarães,
Nuno A. Fonseca, Vítor Santos Costa, Miguel de Sousa,
and Alexandre Magalhães

1 Introduction

Proteins are complex structures synthesised by living organisms. They are actually a fundamental type of molecules and can perform a large number of functions in cell biology. Proteins can assume catalytic roles and accelerate or inhibit chemical reactions in our body. They can assume roles of transportation of smaller molecules, storage, movement, mechanical support, immunity and control of cell growth and differentiation [25]. All of these functions rely on the 3D-structure of the protein. The process of going from a linear sequence of amino acids, that together compose a protein, to the protein's 3D shape is named *protein folding*. Anfinsen's work [29] has proven that primary structure determines the way protein folds. Protein folding is so important that whenever it does not occur correctly it may produce diseases such as Alzheimer's, Bovine Spongiform Encephalopathy (BSE), usually known as *mad cows disease*, Creutzfeldt-Jakob (CJD) disease, a Amyotrophic Lateral Sclerosis (ALS), Huntingtons syndrome, Parkinson disease, and other diseases related to cancer.

A major challenge in Molecular Biology is to unveil the process of protein folding. Several projects have been set up with that purpose. Although protein function is ultimately determined by their 3D structure there have been identified a set of other intermediate structures that can help in the formation of the

Rui Camacho · Rita Ferreira · Natacha Rosa · Vânia Guimarães
LIAAD & Faculdade de Engenharia da Universidade do Porto, Portugal

Nuno A. Fonseca · Vítor Santos Costa
CRACS-INESC Porto LA, Portugal

Vítor Santos Costa
DCC-Faculdade de Ciências da Universidade do Porto, Portugal

Miguel de Sousa · Alexandre Magalhães
REQUIMTE/Faculdade de Ciências da Universidade do Porto, Portugal

M.P. Rocha et al. (Eds.): IWPACBB 2010, AISC 74, pp. 33–41, 2010.
springerlink.com   © Springer-Verlag Berlin Heidelberg 2010

3D structure. We refer the reader to Section 2 for a more detailed description of protein structure. To understand the high complexity of protein folding it is usual to follow a sequence of steps. One starts by identifying the sequence of amino acids (or residues) that compose the protein, the so-called *primary structure*; then we identify the *secondary structures* made of α-helices and β-sheet; and then we predict the *tertiary structure* or 3D shape.

In this paper we address the step of predicting α-helices (parts of the secondary structure) based on the sequence of amino acids that compose a protein. More specifically, in this study we have built models to predict the start of α-helices. We have applied Machine Learning to construct such models. We have collected the sequences of 1499 proteins from the PDB and have assembled data sets that were further used by Machine Learning algorithms to construct the models. We have applied rule induction algorithms, decision trees, functional trees, Bayesian methods, and ensemble methods. We have achieved a 84.4% accuracy and were able to construct some small and intelligible models.

The rest of the paper is organised as follows. Section 2 gives basic definitions on proteins required to understand the reported work. Related work is reported in Section 3. Our experiments, together with the results obtained, are presented in Section 4. Conclusions are presented in Section 5.

2 Proteins

Proteins are build up of amino acids, connect by peptide bonds between the carboxyl and amino groups of adjacent amino acid residues as shown in Figure 1b) [24]. All amino acids have common structural characteristics that include an α carbon to which are connected an amino group and a carboxyl group, an hydrogen and a variable side chain as shown in Figure 1 a). It is the side chain that determines the identity a specific amino acid. There are 20 different amino acids that integrate proteins in cells. Once the amino acids are connected in the protein chain they are designated as residues.

Fig. 1 a) General Structure of an amino acid; side chain is represented by the letter R. b) A fraction of a proteic chain, showing the peptide bounds

In order to function in an organism a protein has to assume a certain 3D conformation. To achieve those conformations apart from the peptide bonds there have to be extra types of weaker bonds between side chains. These extra bonds are responsible for the secondary and tertiary structure of a protein [28].

One can identify four types of structures in a protein. The primary structure
of a protein corresponds to the linear sequence of residues. The secondary structure is composed by subsets of residues arranged as α-helices and β-sheets, as seen
in Figure 2. The tertiary structure results
for the folding of α-helices or β-sheets.
The quaternary structure results from the
interaction of two or more polypeptide
chains.

Fig. 2 Secondary structure conformations of a protein: α-helices
(left); β-sheet (right)

Secondary structures, α-helices and β-sheets, were discovered in 1951 by
Linus Carl Pauling. These secondary structures are obtained due to the flexibility of the peptide chain that can rotate over three different chemical bonds.
Most of the existing proteins have approximately 70% of their structure as
helices that is the most common type of secondary structure.

3   Related Work

Arguably, protein structure prediction is a fundamental problem in Bioinformatics. Early work by Chou et al. [26], based on single residue statistics,
looked for contiguous regions of residues that have an high probability of
belonging to a secondary structure. The protein samples used was very small
which resulted in an overestimation in accuracy of the reported study.

Qian et al [23] used neural networks to predict secondary structures but
achieved an accuracy of only 64.3%. They used a window (of size 13) technique where the secondary structure of the central residues was predicted on
the base of its 12 neighbours.

Rost and Sanderwith used the PHD [3] method on the RS126 data set and
achieved an accuracy of 73.5%. JPRED [14], exploiting multiple sequence alignments, got an accuracy of 72.9%. NNSSP [1] is a scored nearest neighbour
method by considering position of N and C terminal in α-helices and β-strands.
Its prediction accuracy for RS126 data set achieved 72.7%. PREDATOR [7]
used propensity values for seven secondary structures and local sequence alignment. The prediction accuracy of this method for RS126 data set achieved
70.3%. PSIPRED [8] used a position-specific scoring matrix generated by PSI-
BLAST to predict protein secondary structure and achieved 78.3. DSC [18]
used amino acid profile, conservation weights, indels, hydrophobicity were
exploited to achieve 71.1% prediction accuracy in the RS126 data set.

Using a Inductive Logic Programming (ILP) another series of studies
improved the secondary structure prediction score. In 1990 Muggleton et
al. [21] used only 16 proteins (in contrast with 1499 used in our study)
and the GOLEM [22] ILP system to predict if a given residue in a given
position belongs or not to an α-helix. They achieved an accuracy of 81%.

Previous results have been reported by [9] using Neural Networks but achieving only 75% accuracy. The propositional learner PROMIS[17, 30] achieved 73% accuracy on the GOLEM's data set.

It has been shown that the helical occurrence of the 20 type of residues is highly dependent on the location, with a clear distinction between N-terminal, C-terminal and interior positions [16]. The computation of amino acid propensities may be a valuable information both for pre-processing the data and for assessing the quality of the constructed models [10]. According to Blader et al. [4] an important influencing factor in the propensity to form α-helices is the hydrophobicity of the side-chain. Hydrophobic surfaces turn into the inside of the chain giving a strong contribution to the formation of α-helices. It is also known that the protein surrounding environment has influence in the formation of α-helices. Modelling the influence of the environment in the formation of α-helices, although important, is very complex from a data analysis point of view [19].

4 Experiments

4.1 Experimental Settings

To construct models to predict the start of α-helices we have proceeded as follows. We first downloaded a list of low homology proteins from the Dunbrak web site [12][1]. The downloaded list contained 1499 low homology proteins. We then downloaded the PDBs[2] for each of the protein in the list. Each PDB was then processed in order to extract secondary structure information and the linear sequence of residues of the protein.

Each example of a data set is a sequence of a fixed number of residues (window) before and after the start or end of a secondary structure. The window size is fixed for each data set and we have produced 4 data sets using 4 different window sizes. To obtain the example sequences to use we selected sequences such that they are:

1. at the start of a α-helix;
2. at the end of a α-helix;
3. in the interior of a α-helix;
4. at the start of a β-strand;
5. at the end of a β-strand;
6. in the interior of a β-helix.

To do so, we identify the "special" point where the secondary structures start or end, and then add W residues before and after that point. Therefore the sequences are of size $2 \times W + 1$, where $W \in [2, 3, 4, 5]$. In the interior of a secondary structure we just pick sequences of $2 \times W + 1$ residues that do not overlap. With these sequences we envisage to study the start, interior and

---

[1] http://dunbrack.fccc.edu/Guoli/PISCES.php
[2] http://www.rcsb.org/pdb/home/home.do

end of secondary structures. In this paper, however, we just address the first
step of the study, namely, we focus on the start of α-helices.

The size of the data sets, for the different window sizes, are shown in
Table 1.

Table 1 Characterisation of the four data sets according to the window size

Window size	2	3	4	5
Data set size	62053	49243	40529	34337
Number of attributes	253	417	581	745

The attributes used to characterise the examples are of two main types:
whole structure attributes; and, window-based attributes. The whole structure attributes include: the size of the structure; the percentage of hydrophobic residues in the structure; the percentage of polar residues in the structure;
the average value of the hydrophobic degree; the average value of the hy-
drophilic degree; the average volume of the residues; the average area of the
residues in the structure; the average mass of the residues in the structure;
the average isoelectric point of the residues; and, the average topological polar surface area. For the window-based attributes we have used a window of
size W before the "special" point (start or end of either a helix or strand), the
"special" point and a window of size W after the "special" point. For each of
these regions, whenever appropriate, we have computed a set of properties
based on the set of individual properties of residues listed in Table 2.

Table 2 List of amino acid properties used in the study

polarity	hydrophobicity	size	isoelectricpt
charge	h-bonddonor	xlogp3	side chain polarity
acidity	rotatable bond count	h-bondacceptor	side chain charge

For each amino acid of the window and amino acid property we compute
other attributes, namely: the value of the property of each residue in the
window; either if the property "increases" or decreases the value along the
window; the number of residues in the window with a specified value and;
whether a residue at each position of the window belongs to a pre-computed
set of values. Altogether there are between 253 (window size of 2) to 745
(window size of 5) attributes. We have used boolean values: a sequence includes the start of an helix; the sequence does not contain a start of an helix.
All collected sequences where an helix does not start were included in the
"nonStartHelix" class. These later sequences include interior of α-helices, end
points of α-helices, start, interior and end points of beta strands.

The experiments were done in a machine with 2 quad-core Xeon 2.4GHz
and 32 GB of RAM, running Ubuntu 8.10. We used machine learning

algorithms from the Weka 3.6.0 toolkit [31]. We used a 10-fold cross validation procedure to estimate the quality of constructed models. We have used
rule induction algorithms (Ridor), decision trees (J48 [27] and ADTree [11]),
functional trees (FT [13][20]), instance-based learning (IBk [2]), bayesian algorithms (NaiveBayes and BayesNet [15]) and an ensemble method (Random-
Forest [5]).

4.2   Experimental Results

The results obtained with the Machine Learning algorithms are resumed in
Table 3. Apart from the Bayesian methods, most algorithms achieved an accuracy value above the ZeroR predictions. The ZeroR algorithm is used here
as the baseline predictor, it just predicts the majority class. The algorithm
that achieved the best accuracy values was RandomForest, that is an ensemble method. Basically RandomForest constructs several CART-like trees [6]
and produces its prediction by combining the prediction of the constructed
trees.

   For some data mining applications having a very high accuracy is not
enough. In some applications it would be very helpful one can extract knowledge that helps in the understanding of the underlying phenomena that produced the data. That is very true for most of Biological problems addressed
using data mining techniques. In the problem at hands in this paper we
have algorithms that can produce models that are intelligible to experts. J48
and Ridor are examples of such algorithms. Using J48 we mange to produce
a small size decision tree (shown in Figure 3) that uses very informative
attributes near the root of the tree.

Table 3 Accuracy results (%) of the different algorithms on data sets with windows
of size 2, 3, 4 and 5 residues before and after helix start

Algorithm	Window size			
	2	3	4	5
Ridor	83.4	80.6	76.1	77.3
J48	83.9	81.1	79.4	77.0
RandomForest	84.4	81.6	78.4	77.1
FT	79.9	80.5	80.2	75.5
ADTree	83.4	80.3	75.1	76.1
IBk	81.5	76.1	75.2	70.4
NaiveBayes	71.1	66.1	63.2	62.9
BayesNet	70.3	66.2	64.2	64.0
ZeroR	81.5	76.9	72.4	67.8

```
criticalPointSize = tiny
| nHydroHydrophilicWb2 ≤ 1
|      | xlogp3AtPositionA2 ≤ -1.5: noStart (3246.0/816.0)
|      | xlogp3AtPositionA2 > -1.5: helixStart (51.0/24.0)
| nHydroHydrophilicWb2 > 1
|      | rotatablebondcountAtPositionB1 ≤ 1
...
|      | rotatablebondcountAtPositionB1 > 1
...
criticalPointSize = small
| criticalPtGroup = polarweak
|      | chargeAtPositionGroupA2 = negativeneutral: helixStart (1778.0/390.0)
|      | chargeAtPositionGroupA2 = neutralpositive
...
| criticalPointGroup = nonpolarweak: helixStart (1042.0/35.0)
criticalPointSize = large
| chargeAtPositionGroupA2 = negativeneutral
|      | sizeAtPositionGroupB1 = tinysmall
...
|      | sizeAtPositionGroupB1 = smalllarge
...
```

Fig. 3 Attributes tested near the root of a 139 node tree constructed by J48

5   Conclusions and Future Work

In this paper we have addressed a very relevant problem in Molecular Biology, namely that of predicting when, in a sequence of amino acids, an α-helix will start forming. To study this problem we have collected sequences of amino acids from proteins described in the PDB. We have defined two class values: a class of sequences were an α-helix starts forming and; all other types of sequences where an α-helix does not start.

We have applied a set of Machine Learning algorithms and almost all of them made predictions above the naive procedure of predicting the majority class. We have achieved a maximum score of 84.4% accuracy with an ensemble algorithm called Random Forest. We have also managed to construct a small decision tree that has smaller accuracy than 80%, but that is an intelligible model that can help in unveiling the chemical justification of the formation of α-helices.

Acknowledgements

This work has been partially supported by the project ILP-Web-Service (PTDC-/EIA/70841/2006), project STAMPA (PTDC/EIA/67738/2006), and by the Fundação para a Ciência e Tecnologia.

References

1. Salamov, A.A., Solovyev, V.V.: Prediction of protein structure by combining nearest-neighbor algorithms and multiple sequence alignments. J.Mol Biol 247, 11–15 (1995)
2. Aha, D., Kibler, D.: Instance-based learning algorithms. Machine Learning 6, 37–66 (1991)

3. Rost, B.: Phd: predicting 1d protein structure by profile based neural networks. Meth.in Enzym. 266, 525–539 (1996)
4. Blader, M., Zhang, X., Matthews, B.: Structural basis of aminoacid alpha helix propensity. Science 11, 1637–1640 (1993),
   http://www.ncbi.nlm.nih.gov/pubmed/850300
5. Breiman, L.: Random forests. Machine Learning 45(2), 5–32 (2001)
6. Breiman, L., Friedman, J.H., Olshen, R.A., Stone, C.J.: Classification and Regression Trees. Wadsworth International Group, Belmont (1984)
7. Frishman., D., Argos, P.: Seventy-five percent accuracy in protein secondary structure prediction. Proteins 27, 329–335 (1997)
8. Jones, T.D.: Protein secondary structure prediction based on position-specific scoring matrices. Journal of Molecular Biology 292, 195–202 (1999)
9. Kneller, D., Cohen, F.E., Langridge, R.: Improvements in protein secondary structure prediction by an enhanced neural network. Journal of Molecular Biology 216, 441–457 (1990)
10. Fonseca, N., Camacho, R., aes, A.M.: A study on amino acid pairing at the n- and c-termini of helical segments in proteins. PROTEINS: Structure, Function, and Bioinformatics 70(1), 188–196 (2008)
11. Freund, Y., Mason, L.: The alternating decision tree learning algorithm. In: Proceeding of the Sixteenth International Conference on Machine Learning, Bled, Slovenia, pp. 124–133 (1999)
12. Wang, G., Dunbrack Jr., R.L.: Pisces: a protein sequence culling server. Bioinformatics 19, 1589–1591 (2003)
13. Gama, J.: Functional trees. Machine Learning 55(3), 219–250 (2004)
14. Cuff, J.A., Clamp, M.E., Siddiqui, A.S., Finlay, M., Barton, J.G., Sternberg, M.J.E.: Jpred: a consensus secondary structure prediction server. J. Bioinformatics 14(10), 892–893 (1998)
15. John, G.H., Langley, P.: Estimating continuous distributions in bayesian classifiers. In: Eleventh Conference on Uncertainty in Artificial Intelligence, pp. 338–345. Morgan Kaufmann, San Mateo (1995)
16. Richardson, J., Richardson, D.C.: Amino acid preferences for specific locations at the ends of α-helices. Science 240, 1648–1652 (1988)
17. King, R., Sternberg, M.: A machine learning approach for the protein secondary structure. Journal of Molecular Biology 214, 171–182 (1990)
18. King, R., Sternberg, M.: Identification and application of the concepts important for accurate and reliable protein secondary structure prediction. Protein Sci. 5, 2298–2310 (1996)
19. Krittanai, C., Johnson, W.C.: The relative order of helical propensity of amino acids changes with solvent environment. Proteins: Structure, Function, and Genetics 39(2), 132–141 (2000)
20. Landwehr, N., Hall, M., Frank, E.: Logistic model trees. Machine Learning 95(1-2), 161–205 (2005)
21. Muggleton, S. (ed.): Inductive Logic Programming. Academic Press, London (1992)
22. Muggleton, S., Feng, C.: Efficient induction of logic programs. In: Proceedings of the First Conference on Algorithmic Learning Theory, Ohmsha, Tokyo (1990)
23. Qian, N., Sejnowski, T.J.: Predicting the secondary structure of globular proteins using neural network models. Journal of Molecular Biology 202, 865–884 (1988)

24. Petsko, G.A., Petsko, G.A.: Protein Stucture and Function (Primers in Biology). New Science Press Ltd. (2007)
25. Pietzsch, J.: The importance of protein folding. Horizon Symposia (2009), http://www.nature.com/horizon/proteinfolding/background/importance.html
26. Chou, P.Y., Fasman, G.D.: Prediction of secondary structure of proteins from their amino acid sequence. Advances in Enzymology and Related Areas of Molecular Biology 47, 45–148 (1978)
27. Quinlan, R.: C4.5: Programs for Machine Learning. Morgan Kaufmann Publishers, San Mateo (1993)
28. Saraiva, L., Lopes, L.: Universidade Nova de Lisboa, Instituto de Tecnologia Química e Biológica (2007), http://www.cienciaviva.pt/docs/cozinha12.pdf
29. Sela, M., White, F.H., Anfinsen, C.B.: Reductive cleavage of disulfide bridges in ribonuclease. Science 125, 691–692 (1957)
30. Sternberg, M., Lewis, R., King, R., Muggleton, S.: Modelling the structure and function of enzymes by machine learning. Proceedings of the Royal Society of Chemistry: Faraday Discussions 93, 269–280 (1992)
31. Witten, I.H., Frank, E.: Data Mining: Practical machine learning tools and techniques, 2nd edn. Morgan Kaufmann, San Francisco (2005)

A Data Mining Approach for the Detection of High-Risk Breast Cancer Groups

Orlando Anunciação, Bruno C. Gomes, Susana Vinga, Jorge Gaspar,
Arlindo L. Oliveira, and José Rueff

Abstract. It is widely agreed that complex diseases are typically caused by the joint effects of multiple instead of a single genetic variation. These genetic variations may show very little effect individually but strong effect if they occur jointly, a phenomenon known as epistasis or multilocus interaction. In this work, we explore the applicability of decision trees to this problem. A case-control study was performed, composed of 164 controls and 94 cases with 32 SNPs available from the BRCA1, BRCA2 and TP53 genes. There was also information about tobacco and alcohol consumption. We used a Decision Tree to find a group with high-susceptibility of suffering from breast cancer. Our goal was to find one or more leaves with a high percentage of cases and small percentage of controls. To statistically validate the association found, permutation tests were used. We found a high-risk breast cancer

Orlando Anunciação
IST/INESC-ID, TU Lisbon, Portugal
e-mail: orlando@kdbio.inesc-id.pt

Bruno C. Gomes
DG/FCM-UNL, Portugal
e-mail: bruno.gomes@fcm.unl.pt

Susana Vinga
INESC-ID/FCM-UNL Lisbon, Portugal
e-mail: svinga@kdbio.inesc-id.pt

Jorge Gaspar
DG/FCM-UNL, Portugal
e-mail: jgaspar.gene@fcm.unl.pt

Arlindo L. Oliveira
IST/INESC-ID, TU Lisbon, Portugal
e-mail: aml@inesc-id.pt

José Rueff
DG/FCM-UNL, Portugal
e-mail: rueff.gene@fcm.unl.pt

M.P. Rocha et al. (Eds.): IWPACBB 2010, AISC 74, pp. 43–51, 2010.
springerlink.com © Springer-Verlag Berlin Heidelberg 2010

group composed of 13 cases and only 1 control, with a Fisher Exact Test value of 9.7×10^{-6}. After running 10000 permutation tests we obtained a p-value of 0.017. These results show that it is possible to find statistically significant associations with breast cancer by deriving a decision tree and selecting the best leaf.

1 Introduction

Association studies consist on testing the association between markers (called single-nucleotide polymorphisms - SNPs) that are reasonably polymorphic and a phenotype (for example, a disease). It has been pointed out that there is no single haplotype for disease risk and no single protective haplotype but, rather, a collection of haplotypes that confer a graded risk of disease [7]. Probably there is a graded influence of genetic variability in gene expression, whose control depends on many elements [7].

In an association study it is necessary to consider interactions between genetic variations. It is widely agreed that diseases such as ashtma, cancer, diabetes, hypertension and obesity are typically caused by the joint effects of multiple genetic variations instead of a single genetic variation [17] [23]. These multiple genetic variations may show very little effect individually but strong interactions jointly, a phenomenon known as epistasis or multilocus interaction [3] [26].

Methods for the detection of epistatic interactions are very important since if the effect of one locus is altered or masked by effects at another locus, power to detect the first locus is likely to be reduced and elucidation of the joint effects at the two loci will be hindered by their interaction [3]. Recently, an increasing number of reasearchers have reported the presence of epistatic interactions in complex diseases [5] [26], such as breast cancer [20] and type-2 diabetes [2]. Multifactor-Dimensionality Reduction (MDR) method was used to find a four-locus interaction associated with breast cancer [20]. MDR was also applied to find a two-locus interaction associated with a reduced risk of type-2 diabetes [2]. Exhaustive search methods such as MDR work well on small size problem. However, in Genome-Wide Association studies, direct application of these methods is computationally prohibitive [26].

A number of different methods have been used, including statistical methods (e.g. ATOM [10]), search methods (e.g. BEAM [28]), regression methods (e.g. Lasso Penalized Logistic Regression [24]) and machine learning methods (e.g. [11] or MegaSNPHunter [25]). However, there is a need for new tools to accurately discover the relationship between combinations of SNPs, other genetic variations and environmental exposure, with disease susceptibility in the context of Genome-Wide Association Studies [15]. In this paper, we explore the applicability of decision tree learning to a breast cancer association study.

In this work we are not trying to find a classifier for distinguishing between all types of cases and controls. We are focused on finding a classifier that maximizes the number of cases and minimizes the number of false positives (controls). We want a classifier with low probability of producing false positives. Such classifier will miss many cases (high number of false negatives) but will have high accuracy when someone is classified as a case (low number of false positives). Decision trees have the ability to produce such a classifier if we select a leaf with few controls and many cases. Decision tree inducers select at each step the variable that best separates cases from controls, according to a certain splitting criterion (see Section 2.2 for more details), and are therefore suitable to find groups with low number of false positives. On the other hand, decision tree induction is a greedy algorithm and thus can have problems in the detection of multilocus interactions in which each individual attribute is not associated with the disease (that is interactions with low marginals). However, our results show that, even with this limitation, it is possible to find a statistically significant association with the disease.

The ultimate goal of association studies is to facilitate a systems-based understanding of disease, in which we come to understand the full, molecular network that is perturbed in disease [7]. The discovery of statistical interaction does not necessarily imply interaction on the biological or mechanistic level. However, allowing for different modes of interaction between potential disease loci can lead to improved power for detection of genetic effects. We may, therefore, succeed in identifying genetic variants that might otherwise have remained undetected [3].

2 Methods

In this section we present a brief description of the methods that were used and describe our experimental procedure.

2.1 Data Description

We used a dataset produced by the Department of Genetics of the Faculty of Medical Sciences of Universidade Nova de Lisboa with 164 controls and 94 cases, all of them being portuguese caucasians. Of the 94 cases, 50 of them had its tumour detected after menopause in women above 60 years old, while the other 44 had its tumour detected before menopause, in women under 50 years old. The tumour type is ductal carcinoma (invasive and in situ). SNPs were selected with Minor Allele Frequency above or equal to 5% for european caucasian population (HapMap CEU). Tag SNPs were selected with a correlation coefficient $r^2 = 0.8$. A total of 32 SNPs are available, 7 from the BRCA1 gene (rs16942, rs4986850, rs799923, rs3737559, rs8176091, rs8176199 and rs817619), 19 from the BRCA2 gene (rs1801406, rs543304, rs144848, rs28897729, rs28897758, rs15869, rs11571836, rs1799943, rs206118, rs2126042, rs542551, rs206079, rs11571590, rs206119, rs9562605, rs11571686, rs11571789, rs2238163 and rs1012130) and 6 from the TP53 gene (rs1042522, rs8064946, rs8079544, rs12602273, rs12951053 and rs1625895). SNPs belong to several parts of the gene: regulatory region, coding region or non-coding region. The genotyping was done with real time PCR (Taqman technology). Tobacco and alcohol consumption were also used as attributes for the analysis.

2.2 Decision Trees

Decision Tree Learning is one of the most widely used and practical methods for classification [14]. In this method, learned trees can be represented as a set of if-then rules that improve human readability. Decision trees are very simple to understand and interpret by domain experts.

A decision tree consists of nodes that have exactly one incoming edge, except the root node that has no incoming edges. A node with outgoing edges is an internal (or test) node, while the other nodes are called leaves (also known as terminal nodes or decision nodes).

Each internal node splits the instance space into two or more subspaces, according to a discrete function on the input attributes values. Usually each discrete function is a test that considers a single attribute and its corresponding value. Each leaf is associated with one class, representing a value of the target variable given the values of the variables represented by the path from the root node. The classification of one instance is done by navigating from the root node down to a leaf, according to the outcome of the tests along the path. For this to be possible, a decision tree must cover the space of all possible instances. In a decision tree each node is labeled with the attribute it tests, and its branches are labeled with its corresponding values.

Less complex decision trees increase model interpretability for domain experts. However, the tree complexity has a crucial effect on its accuracy [1]. The tree complexity is explicitly controlled by the stopping criteria used and the pruning method employed.

To build a decision tree from a given dataset an algorithm called a decision tree inducer is used. There are several inducers such as ID3 [18], C4.5 [19] and CART [1].

In general, inducing a minimal decision tree consistent with the training set is NP-Hard [6]. It was also shown that building a minimal binary tree with respect to the expected number of tests required for classifying an unseen instance is NP-complete [8]. Even the problem of finding the minimal equivalent decision tree for a given decision tree is NP-Hard [27].

Since building optimal decision trees is unfeasible in real problems, induction algorithms must use heuristics. Most algorithms use recursive top-down approaches which partition the training space. These algorithms are greedy since at each step the best split is chosen (according to some splitting criterion) [21].

The splitting criterion is the criterion that is used for selecting the best split at each step of the inducer algorithm. Splitting criteria can be univariate, which means that only one attribute is used for partition the space at each step of the algorithm, or multivariate in which multiple attributes are considered at each step. There are several multivariate splitting criteria [1] [4] [16]. However, univariate splitting criteria are simpler and more popular [21]. There are several univariate splitting criteria such as Information Gain, Gain Ratio or Gini Index. In most cases, the choice of splitting criteria will not make much difference on the tree performance [21].

Deciding when to stop building a decision tree is not an easy task. Stopping too early may result in small and under-fitted decision trees. Stopping too late results in overfitted decision trees. The typical way to deal with this problem is to create an overfitted decision tree and then perform pruning. There are several methods for pruning of decision trees such as Reduced Error Pruning, Minimum-Error Pruning or Minimum Description Length Pruning [13].

2.3 Statistical Validation

In this subsection we present the methods used to statistically validate our results.

2.3.1 Fisher Exact Test

Fisher's exact test (FET) is a statistical test used to determine if there are nonrandom associations between two categorical variables [22].

In the case of a 2×2 matrix, the conditional probability of getting the actual matrix given the particular row and column sums is given by Equation 1.

$$P_{\text{cutoff}} = \frac{(a+b)!(c+d)!(a+c)!(b+d)!}{n!a!b!c!d!} \qquad (1)$$

In Equation 1 a, b, c and d are the 4 entries of the 2×2 matrix and $n = a+b+c+d$. If we want to calculate the p-value of the test, we can do it by computing the sum of all p-values which are $\leq P_{\text{cutoff}}$.

To compute the value of Fisher Exact Test given by Equation 1, we need to decompose it in order to avoid computing large factorials. An equivalent formulation of Equation 1 is given by:

$$P_{\text{cutoff}} = e^{s(a+b)+s(c+d)+s(a+c)+s(b+d)-s(n)-s(a)-s(b)-s(c)-s(d)} \qquad (2)$$

in which $s(m) = \sum_{i=1}^{m} \log i$.

This test will be applied to evaluate the quality of leaves in the decision tree as described in Section 3.

2.3.2 Permutation Tests

Permutation tests are non-parametric procedures for determining statistical significance based on rearrangements of the labels of a dataset. It is a robust method, but it can be computationally intensive. A test statistic, which is computed from the dataset, is compared with the distribution of permutation values. These permutation values are computed similarly to the test statistic, but, under a random rearrangement of the labels of the dataset [9]. Permutation tests can help to reduce the multiple testing burden [12] and can be used to compare statistical tests [10].

In bioinformatics, permutation tests have become a widely used technique. The reason for this popularity has to do with its non-parametric nature, since in many bioinformatics applications there is no solid evidence or sufficient data to assume a particular model for the obtained measurements of the biological events under investigation [9].

The main disadvantage of permutation testing is that it needs a very large number of permutations when small p-values are to be accurately estimated, which is computationally expensive. To address this problem, the tail of the distribution of permutation values can be approximated by a generalized Pareto distribution [9]. According to [9] accurate P-value estimates can be obtained with a drastically reduced number of permutations when compared with the standard empirical way of computing p-values.

3 Experimental Procedure

In order to find a group with high-susceptibility of suffering from breast cancer, we started by applying decision trees with a 10-fold cross validation strategy. 10-fold cross validation is a strategy that is based on performing 10 iterations. In each of the 10 iterations, a different portion of 1/10 of the dataset is used as an independent test set while the remaining 9/10 of the dataset is used for training.

To conduct our experiments, we used Weka J48 Decision Tree which generates a C4.5 decision tree. Several parameters were tested such as the confidence factor used for pruning, whether to use binary splits or not, whether to prune the tree or not and the minimum number of instances per leaf. For each different combination of parameters, we saved the average classification accuracy of the 10 folds. We then selected as the best combination of parameters, the combination that had a higher average classification accuracy on 10-fold cross validation. We used the C4.5 decision tree inducer algorithm with the best combination of parameters using the entire dataset for training in order to build our final model.

After our final model is built, we looked into the tree, selected the best leaf L according to Fisher Exact Test value (lower values of FET mean stronger associations) and saved this value, $FET(L)$.

In order to statistically validate our detected association, we used 10000 permutation tests. We state as our null hypothesis H_0 that there is no association between our variables and the phenotype. If we find a strong association between one or more variables and the phenotype, then we can reject our null hypothesis. In order to test whether our null hypothesis can be rejected, for each permutation test we randomly permute labels (phenotype attribute). We then build a new decision tree with the same parameters as in our original dataset. We check all leaves from the decision tree and save the lowest FET value. This means that in each permutation test we save the FET value of the best leaf. In the end of the execution of the permutation tests, we have 10000 FET values that were generated according to the underlying and unknown distribution of the null hypothesis. We can then check where the FET value of our leaf L stands on the estimated distribution of our null hypothesis. This way we can see if $FET(L)$ is extreme enough so that we can reject the null hypothesis with high statistical confidence, therefore obtaining our adjusted p-value.

4 Results

Figure 1 shows the decision tree learned over the entire dataset using the best parameters found in 10-fold cross validation. We can see that the attribute on the root

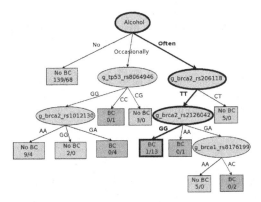

Fig. 1 Decision Tree Model learned on our dataset. In the leaves there is information about the controls/cases distribution

of the tree is alcohol. On the questionnaires, people answered if they drink alcohol with meals (often), socially (occasionally) or if they do not drink alcohol at all (no). There is a very interesting leaf L in this tree: **Alcohol=often and rs206118=TT and rs2126042=GG**. The group of people that has this combination of factors is composed by 13 cases and only 1 control. This gives us a $\text{FET}(L) = 9.7 \times 10^{-6}$. The adjusted p-value obtained with 10000 permutation tests as described in Section 3 was 0.017, which means that we can reject our null hypothesis at the 2% level of confidence.

5 Conclusions and Future Work

With this methodology, we showed that it is possible to find statistically significant associations from a breast cancer data set. However, this methodology needs to be evaluated in a larger set of examples in order to find associations with a higher degree of statistical confidence. Using a larger data set will also enable us to find correlations between a bigger set of genes and SNPs. We have to be very careful when trying to extract biological knowledge from the statistical conclusions. In this work we found an association from leaf L that involves alcohol consumption and two SNPs from BRCA2 gene. To extract biological knowledge from this detected association we have to take into account that SNPs were selected for genotyping using a tag SNP selection method with a correlation coefficient $r^2 = 0.8$. This means that if this detected association has an underlying biological process that supports it, we cannot say which SNPs are involved: if SNPs rs206118 and rs2126042 or other SNPs that are in linkage disequilibrium with these two. This means that to extract biological knowledge from a statistically significant association we will need expert knowledge about biological processes related to breast cancer.

We also have to take into account that decision tree learning is a greedy algorithm. It selects at each step the SNP that maximizes the splitting criterion used which has the problem that interactions in which each individual attribute is not associated with the disease (that is interactions with low marginals) are not detected. The algorithm has to be modified in order to be able to also detect interactions with low marginals.

Acknowledgements. The authors acknowledge financial support from Fundação para a Ciência e Tecnologia (SFRH / BD / 41984 / 2007) and project SHIPs: Sat-based Haplotype Inference by Pure Parsimon (FCT Project PTDC/EIA/64164/2006).

References

1. Breiman, L., Friedman, J., Olshen, R., Stone, C.: Classification and regression trees. Wadsworth, Belmont (1984)
2. Cho, Y.M., Ritchie, M.D., Moore, J.H., Park, J.Y., Lee, K.U., Shin, H.D., Lee, H.K., Park, K.S.: Multifactor-dimensionality reduction shows a two-locus interaction associated with Type 2 diabetes mellitus. Diabetologia 47(3), 549–554 (2004)
3. Cordell, H.J.: Epistasis: what it means, what it doesn't mean, and statistical methods to detect it in humans. Human Molecular Genetics 11(20), 2463–2468 (2002)
4. Duda, R., Hart, P.: Pattern Classification and Scene Analysis. Wiley, New York (1973)
5. Griffiths, A.J.F., Wessler, S.R., Lewontin, R.C., Gelbart, W.M., Suzuki, D.T., Miller, J.H.: Introduction to Genetic Analysis. W.H. Freeman and Co Ltd., New York (2008)
6. Hancock, T.R., Jiang, T., Li, M., Tromp, J.: Lower bounds on learning decision lists and trees. Inform. Comput. 126(2), 114–122 (1996)
7. Hardy, J., Singleton, A.: Genomewide association studies and human disease. New England Journal of Medicine 360(17), 1759–1768 (2009)
8. Hyafil, L., Rivest, R.L.: Constructing optimal binary decision trees is np-complete. Inform. Process. Lett. 5(1), 15–17 (1976)
9. Knijnenburg, T.A., Wessels, L.F., Reinders, M.J., Shmulevich, I.: Fewer permutations, more accurate P-values. In: Bioinformatics, vol. 25(ISMB 2009), pp. i161–i168 (2009)
10. Li, M., Wang, K., Grant, S.F.A., Hakonarson, H., Li, C.: ATOM: a powerful gene-based association test by combining optimally weighted markers. Bioinformatics 25(4), 497 (2009)
11. Listgarten, J., Damaraju, S., Poulin, B., Cook, L., Dufour, J., Driga, A., Mackey, J., Wishart, D., Greiner, R., Zanke, B.: Predictive models for breast cancer susceptibility from multiple single nucleotide polymorphisms. Clinical Cancer Research 10, 2725–2737 (2004)
12. Marchini, J., Donnelly, P., Cardon, L.R.: Genome-wide strategies for detecting multiple loci that influence complex diseases. Nature Genetics 37(4), 413–417 (2005)
13. Mehta, R.L., Rissanen, J., Agrawal, R.: Mdl-based decision tree pruning. In: Proc. 1st Int. Conf. Knowledge Discovery and Data Mining, pp. 216–221 (1995)
14. Mitchell, T.M.: Machine Learning. McGraw-Hill, New York (1997)
15. Moore, J.H., Asselbergs, F.W., Williams, S.M.: Bioinformatics Challenges for Genome-Wide Association Studies. Bioinformatics 26(4), 445–455 (2010)
16. Murthy, S.K., Kasif, S., Salzberg, S.: A system for induction of oblique decision trees. J. Artif. Intell. Res. 2, 1–33 (1994)
17. Musani, S.K., Shriner, D., Liu, N., Feng, R., Coffey, C.S., Yi, N., Tiwari, H.K., Allison, D.B.: Detection of gene× gene interactions in genome-wide association studies of human population data. Hum. Hered. 63(2), 67–84 (2007)
18. Quinlan, J.R.: Induction of decision trees. Mach. Learn. 1, 81–106 (1986)

19. Quinlan, J.R.: C4.5: Programs for Machine Learning. Morgan Kaufmann, San Francisco (1993)
20. Ritchie, M.D., Hahn, L.W., Roodi, N., Bailey, L.R., Dupont, W.D., Parl, F.F., Moore, J.H.: Multifactor-dimensionality reduction reveals high-order interactions among estrogen-metabolism genes in sporadic breast cancer. The American Journal of Human Genetics 69(1), 138–147 (2001)
21. Rokach, L., Maimon, O.: Top-down induction of decision trees classifiers - a survey. IEEE Transactions on Systems, Man, and Cybernetics - Part C: Applications and Reviews 35(4), 476–487 (2005)
22. Weisstein, E.W.: Fisher's exact test. MathWorld – A Wolfram Web Resource., http://mathworld.wolfram.com/AffineTransformation.html
23. Wongseree, W., Assawamakin, A., Piroonratana, T., Sinsomros, S., Limwongse, C., Chaiyaratana, N.: Detecting purely epistatic multi-locus interactions by an omnibus permutation test on ensembles of two-locus analyses. BMC bioinformatics 10(1), 294 (2009)
24. Wu, T.T., Chen, Y.F., Hastie, T., Sobel, E., Lange, K.: Genome-wide association analysis by lasso penalized logistic regression. Bioinformatics 25(6), 714–721 (2009)
25. Xiang, W., Can, Y., Qiang, Y., Hong, X., Nelson, T., Weichuan, Y.: MegaSNPHunter: a learning approach to detect disease predisposition SNPs and high level interactions in genome wide association study. BMC Bioinformatics 10(13) (2009)
26. Yang, C., He, Z., Wan, X., Yang, Q., Xue, H., Yu, W.: SNPHarvester: a filtering-based approach for detecting epistatic interactions in genome-wide association studies. Bioinformatics 25(4), 504 (2009)
27. Zantema, H., Bodlaender, H.L.: Finding small equivalent decision trees is hard. Int. J. Found. Comput. Sci. 11(2), 343–354 (2000)
28. Zhang, Y., Liu, J.S.: Bayesian inference of epistatic interactions in case-control studies. Nature genetics 39(9), 1167–1173 (2007)

GRASP for Instance Selection in Medical Data Sets

Alfonso Fernández, Abraham Duarte, Rosa Hernández, and Ángel Sánchez

Abstract. Medical data sets consist of a huge amount of data organized in instances, where each one contains several attributes. The quality of the models obtained from a database strongly depends on the information previously stored on it. For this reason, these data sets must be preprocessed in order to have fairly information about patients. Data sets are preprocessed reducing the amount of data. For this task, we propose a GRASP algorithm with two different improvement strategies based on Tabu Search and Variable Neighborhood Search. Our procedure is able to widely reduce the original data keeping the most relevant information. Experimental results show how our GRASP is able to outperform the state of the art methods.

1 Introduction

Almost every day, medical staff diagnoses whether a patient has a disease or not in order to apply the pertinent treatment. The diagnosis is based on different analyses/tests performed to the patient and the expertise of the doctors. This process could be eased if medical staff were able to find common patterns with other patients that have suffered the same disease. In this way, if it is compared the medical record of the current patient with other patients previously treated, doctors could infer a diagnosis based of the history. However, it means to deal with massive amounts of information and collections of data. A very active line of research focuses on scaling down data, where the main problem is how to select the relevant data. This task is carried out in the data preprocessing phase in a Knowledge Discovery in Databases (KDD) process. The goal of this area consists of withdrawing relevant information from databases using a systematic and detailed analysis of the data [5]. This information is used in Data Mining, DM, to create models useful for science, engineering or economy [8]. As it is reported in the literature, DM models are very dependent on the quality of the stored data. Therefore, the first phase KDD is the preprocessing the original data whose main target is improving the "quality" of data.

Alfonso Fernández · Abraham Duarte · Rosa Hernández · Ángel Sánchez
Departamento de Ciencias de la Computación, URJC
e-mail: alfonso.fernandez@urjc.es, abraham.duarte@urjc.es,
    rm.hernandez@alumnos.urjc.es, angel.sanchez@urjc.es

M.P. Rocha et al. (Eds.): IWPACBB 2010, AISC 74, pp. 53–60, 2010.
springerlink.com   © Springer-Verlag Berlin Heidelberg 2010

In Data Mining there are several preprocessing techniques. Among them, we can highlight *Data Reduction, Data Cleaning, Data Integration* and *Data Transformation*. The reader is referred to [1] to find detailed descriptions of these strategies. In this work, we focus on Data Reduction (DR). It can be achieved in many ways. Specifically, in the literature we can find DR based on selecting features, making the feature-values discrete and selecting instances. This paper is devoted to Instance Selection (IS) as DR mechanism [11]. IS consists of reducing the number of rows in a data set where each row represents an instance. There are several IS strategies. Among them we can highlight sampling, boosting, prototype selection, and active learning. We will study IS from the prototype selection (PS) perspective, called IS-PS.

There are several papers in the literature that have studied this topic. Most of them are based on Evolutionary Strategies. The first relevant paper, presented by Kuncheva [10], is an Evolutionary Algorithm for PS. In [1] is presented PBIL as the first combination of Genetic Algorithm and Competitive Learning designed for searches in binary spaces. Eshelman presented in [3] CHC, considered a reference in the Genetic Algorithm field because it introduces a diversity mechanism to obtain a good balance between intensification and diversification in the search process. Cano et al introduced in [2] a Memetic Algorithm, MA, that solves the scalability problem in prototype selection. Finally, in [6] is presented the most recent work in the context of IS-PS. It is an improved MA, called SSMA, which outperforms previous approaches.

In this work, we have designed a procedure based on several metaheuristics, to preprocess medical data sets. The goal of our procedure consists of reducing the set of original data obtaining a subset of data that fairly represents the original set.

2 Preprocessing in KDD

In general, data sets are arranged on a table, *OriginalTable*, where each row corresponds to an *Instance* and each column to an attribute. Each instance is characterized by a set of attributes and classified in a determined class (according to the values of their corresponding attributes). IS techniques construct a smaller table, *ReducedTable*, selecting the smallest set of instances that enable a given algorithm to predict the class of a query instance with the same (or higher) accuracy as the original set. This reduction improves both space and time complexities of subsequent DM strategies. It is important to remark that removing instances does not necessarily lead to a degradation of the results. This behavior could be explained taking into account that some data with noise or repeated be deleted by removing instances.

The main goal of our proposal consists of constructing the *ReducedTable* $\subset$ *OriginalTable* with the most representative instances and the larger capacity of classifying new instances. In order to classify an instance in the corresponding class, we will use the Nearest Neighborhood (1-NN strategy) as customary. In order to classify a new instance using *ReducedTable*, we compute the distances from this new instance to the rest of instances in *ReducedTable*. Finally, the new instance is assigned to the same class of the nearest neighbor one. As in previous works, we use the Euclidean distance defined in an n-dimensional space, where n represents the number of attributes [2].

To determine the quality of the IS-PS technique we define a *fitness* function that combines two values: the classification performance (*%Clas*) and the percentage of reduction (*%Red*). This function is a tradeoff between the ability of *ReducedTable* to classify instances and the reduction in the data performed for IS-PS. In mathematical terms, the *fitness* function *f* is:

$$f = \alpha*(\%Clas) + (1-\alpha)*(\%Red) \qquad (1)$$

The 1-NN classifier is used for measuring the classification rate, *%Clas*, and denotes the percentage of correctly classified instances and *%Red*, is defined as:

$$\%Red = 100* (|OriginalTable| - |ReducedTable|)/|OriginalTable| \qquad (2)$$

where |*OriginalTable*| is the number of instances in the original table and |*ReducedTable*| is the number of instances in the reduced table. The objective of the proposed algorithm is to maximize the *fitness* function. As a consequence, it is maximized the classification rate and minimized the resulting number of instances. The value of α ranges from 0 to 1. It measures the emphasis given to precision (percentage of classification) and reduction (percentage of reduction). In this work, we set α to 0.5, balancing precision and reduction.

3 GRASP for IS-PS

The GRASP methodology was developed in the late 1980s, and the acronym was coined by Feo and Resende in 1995[4]. Each iteration consists of constructing a trial solution and then applying an improvement procedure to find a local optimum (i.e., the final solution for that iteration). The construction phase is iterative, randomized greedy, and adaptive. In this section we describe our adaptation of the GRASP methodology to IS-PS.

3.1 Constructive Algorithm

The construction phase is greedy, randomized and adaptive. It is greedy because the addition of each element is guided by a greedy function. It is randomized because a random selection takes place and the information provided by the greedy function is used in combination with this random element. Note that the randomness in GRASP allows multiple iterations obtaining different solutions. Finally, it is adaptive because the element chosen at any iteration in a construction is a function of those previously chosen (updating relevant information from one construction step to the next).

In the context of IS-PS, the constructive algorithm starts by computing the center of gravity of a class. The center, $s\_center(X)$, of a set of elements belonging to class $X = \{s_i : i \in I\}$ is defined as:

$$s\_center(X) = \frac{\sum_{i \in I} s_i}{|X|} \tag{3}$$

where I represent the set of different attributes. $s\_center(X)$ is "virtual" instance where the value of each attribute is computed as the average of the attributes of every instance that belongs to X. In order to have a "real" instance instead of a "virtual" instance, we select from the original table the nearest instance to each virtual instance. Therefore, if the data set has m classes, we compute m centers of gravity (one for each class), and we initialize *ReducedTable* with these m real instances. One time we have this table; we can compute the *fitness* as defined above. Obviously, we will have the largest possible percentage of reduction but the percentage of classification is worst.

To simplify the notation, we call *Sel* as the set of instances in *ReducedTable* and *Unsel* as the set of instances in *OriginalTable* – *ReducedTable*. *Sel* contains the set of selected instances and *Unsel* contains the set of unselected instances.

The GRASP constructive procedure improves *%Class* by adding new instances to *Sel* one at a time. In order to do so, it is computed the *fitness f(v)*, for each instance $v \in Unsel$ if instance v were included in *Sel*. Notice that the larger $f(v)$ the better the improvement. This is the greedy part of the algorithm.

All the instances in *Unsel* with a *fitness* value strictly positive are candidates to be included in *Sel*. We call them as Candidate List (*CL*),

$$CL = \{v \in Unsel \,/\, f(v) > 0\} \tag{4}$$

We define the Restricted Candidate List (*RCL*), as the set of elements with larger $f(v)$ values. In mathematical terms:

$$RCL = \{v \in CL \,/\, f(v) \geq f_{th}\} \tag{5}$$

where f_{th} is a threshold computed as a percentage β between the maximum, f_{max}, and minimum, f_{min}), values of the instances in *Unsel*:

$$f_{th} = f_{min} + \beta(f_{max} - f_{min}) \tag{6}$$

where $f_{max} = max\,f(v)\, f_{min} = min\,f(v)$ with $v \in CL$. If $\beta = 1$, the algorithm is completely greedy. On the other hand, if $\beta = 0$ the algorithm is purely random. To favor the "diversification" of the procedure, an instance is randomly selected at each iteration to be included in *Sel*. This is the random part of the algorithm. The inclusion of the new instance in *Sel* yields to a modification of the computation of $f(v)$ and then, to a new *RCL*. This is the adaptive part of the algorithm.

The constructive process is maintained, including instances into *Sel*, until no further improvement in the *fitness* is obtained. Then it stops and returns the constructed solution.

3.2 Local Search

The second phase of our solving method is an improvement procedure. Specifically, we propose a local search, LS, based on removing/adding instances. It means that solutions reachable from the incumbent one are those constructed by

removing or adding one instance to *Sel*. Specifically, LS starts by removing instances from *Sel* until no further improvement in the *fitness* value is obtained.
Then, the local search resort to add new instances, selecting in each iteration an
instance in *Unsel* able to improve the current *fitness*. LS performs insertion
movements while the *fitness* value increases. LS keeps removing/ adding instances
until no further improvement is possible.

3.3 Tabu Search

Tabu Search, TS is a metaheuristic that guides a local search procedure to explore
the solution space beyond local optimality [7]. One of the main components of TS
is its use of adaptive memory, which creates more flexible search behavior.

   The structure of a neighborhood in TS goes beyond that used in local search by
embracing the types of moves used in constructive and destructive processes
(where the foundations for such moves are accordingly called constructive neighborhoods and destructive neighborhoods). We can implement memory structures
to favor (or avoid) the inclusion of certain elements in the solution previously
identified as attractive (or unattractive). Such expanded uses of the neighborhood
concept reinforce a fundamental perspective of TS, which is to define neighborhoods in dynamic ways that can include serial or simultaneous consideration of
multiple types of moves.

   We propose a TS strategy based on LS. In destructive neighborhoods, TS selects the instance x in *Sel* with the lowest contribution, $f(x)$, to the *fitness* function.
It means that we would obtain the best possible *fitness* removing x. On the other
hand, in constructive neighborhoods, TS selects the instance y in *Unsel* which
were able to improve upon the *fitness* value. Every instance involved in a movement becomes tabu for *Tenure* iterations. As it is customary in TS, we permit non-
improving moves that deteriorate the objective value. TS stops after *MaxIterations*
without improving the best found solution.

3.4 Variable Neighborhood Search

The Variable Neighborhood Search (VNS), proposed by Hansen and Mladenovíc
[9], is a metaheuristic whose basic idea is a systematic change of neighborhood
within a local search. Each step in VNS has three major phases: neighbor generation, local search and jump. Unlike to other metaheuristics, VNS allows changes
of the neighborhood structure during the search. VNS explores increasingly
neighborhoods of the current best found solution. The basic idea is to change the
neighborhood structure when the local search is trapped on a local minimum.

   Let N_k, $k = 1,\dots, k_{max}$ be a set of predefined neighborhood structures and let
$N_k(x)$ be the set of solutions in the kth-order neighborhood of a solution x. Specifically the kth-order neighborhood is defined by all solutions that can be derived
from the current one by selecting k instances and transferring each instance from
Sel to *Unsel* and vice versa.

   Our VNS strategy is based on LS, but it generalizes the constructive/destructive
neighborhoods. Specifically, starting from a solution x, VNS select k instances at

random (instead of 1) to be added to/removed from *Sel* obtaining a solution *y*. After that, it applies LS. If the new improved solution has a better *fitness* than the original one, then *k* is set to 1 and the search jumps to the new solution. Otherwise, *k* = *k*+1 and VNS tries again to add/remove *k* instances at random. The procedure stops when *k* reaches a maximum value.

5 Experimental Results

All experiments were conducted on a personal computer with a Pentium IV Core 2 Duo 2.27 GHz with 2 GB RAM. We coded all the procedures in Java and the number of iterations of the GRASP algorithm was set to 20.

In order to evaluate the behavior of the algorithms applied in different size data sets, we have carried out a number of experiments increasing complexity and size of data sets. We have selected seven test sets, which cover a wide size range, as we can see in Table 1. They are available at http://archive.ics.uci.edu/ml/.

Table 1 Medical datasets

Name	Instances	Attributes	Classes
Lymphography	148	18	4
Cleveland	303	14	2
Bupa	345	7	2
Wisconsin	683	9	2
Pima	768	8	2
Splice	6435	36	3
Thyroid	7200	21	3

For example, *Cleveland* data set contains information about patients with or without cardiovascular problems. Some of its attributes are the number of cigarettes smoked a day, the age of the patient or if an antecessor of the family of the patient suffered from similar problems. In *Wisconsin*, the data are gathered from the analysis of different samples of lung tissue of patients that could suffer from lung cancer. In this case, the attributes correspond to measurements of cell morphology such as area, radius or perimeter. *Splice* junctions are points on a DNA sequence at which "superfluous" DNA is removed during the process of protein creation in higher organisms. This problem consists of three subtasks: recognizing exon/intron boundaries (referred to as EI sites), recognizing intron/exon boundaries (IE sites) or none of them.

Table 2 reports the average results for the test sets introduced in Table 1. All the results were computed using a 10-fold cross validation scheme showing the average values. In the first column, CPU Time, we report the average running time. In the second column, *%Class*, is represented the average percentage of classification with its corresponding standard deviation. Finally, in the third column, *%Red*, shows the average percentage of reduction.

Table 2 Results for all the medical datasets

Name	GRASP_TS			GRASP_VNS		
	CPU Time	%Class	%Red	CPU Time	%Class	%Red
Lymphography	6.18	91.67 ± 1.19	90.69 ± 1.35	0.89	90.32 ± 1.09	92.19 ± 0.81
Cleveland	30.9	69.18 ± 1.59	90.42 ± 1.27	7.96	65.74 ± 1.45	93.19 ± 0.83
Bupa	37.8	85.76 ± 1.12	91.40 ± 1.20	5.93	79.97 ±1.46	96.01 ± 0.73
Wisconsin	105.5	98.15 ± 0.28	99.38 ± 0.18	4.75	97.82 ± 0.31	99.61 ± 0.08
Pima	172.2	81.48 ± 1.12	97.11 ± 0.68	45.5	80.93 ± 0.70	98.44 ± 0.21
Splice	12384	92.30 ± 0.46	95.53 ± 0.26	3418	91.05 ± 0.46	96.35 ± 0.26
Thyroid	7695	95.11 ± 0.22	99.50 ± 0.10	3067	94.96 ± 0.12	99.64 ± 0.09

Table 2 shows the merit of the proposed procedures. Our GRASP implementations, GRASP_TS and GRASP_VNS, consistently produce high quality solutions. GRASP_TS marginally improves GRASP_VNS in terms of %Class, while GRASP_VNS improves GRASP_TS in terms of %Red. Summarizing, the behavior of both procedures is quite similar. The robustness of the method fact can be observed in the small value of the standard deviation. Regarding the percentage of reduction, as an example, in the test set Wisconsin, the final *ReducedTable* contains only two instances (one for each class) classifying correctly on average 97.65% of the instances. On the other hand, it is important to remark that our approach is able to reduce the set of instances from thousands (i.e Thyroid) to tens (1%), with a CPU time considerably larger.

Having determined the quality of our algorithm, we compare our GRASP algorithms with the best method identified in previous studies [6]. We employ in each experiment not only the same test sets but also the conditions and evaluation criteria found in the respective papers. Tables 3 shows the *fitness* value for both procedures executed over the whole test set.

Table 3 Results for medium medical datasets

Name	GRASP_TS	GRASP_VNS	SSMA
Lymphography	91.18 ± 0.50	**91.26 ± 0.41**	75.23 ± 1.47
Cleveland	79.80 ± 0.30	79.46 ± 0.45	**80.68 ± 0.82**
Bupa	**88.58 ± 0.38**	87.99 ± 0.54	85.78 ± 1.07
Wisconsin	**98.76 ± 0.14**	98.71 ± 0.15	98.52 ± 0.11
Pima	89.29 ± 0.38	89.68 ± 0.33	**89.77 ± 0.48**
Splice	**93.91 ± 0.14**	93.70 ± 0.14	75.23 ± 1.47
Thyroid	**97.30 ± 0.09**	97.30 ± 0.07	97.08 ± 0.11
Avg. Value	91.26 ± 0.28	91.16 ± 0.29	86.04 ± 0.79

The best values of the *fitness* are bolded. The results in Tables 3 indicate that GRASP is capable of finding high quality solutions, defeating to SSMA in 5 (out of 7) test sets. Regarding the average *fitness* GRASP compares favorably to SSMA in both improvement methods (TS and VNS).

6 Conclusions

We have described the development and implementation of a GRASP algorithm for IS–PS problem. We propose a new constructive procedure based on the computation of the center of gravity. We have also described a new improvement method based on two different types of movements: exchanges and add/remove instances. Additionally, we have proposed two advanced improvement strategies based on the TS and VNS methodologies. We are able to produce a method that reaches good quality solutions on previously reported problems. Our algorithm is compared to state-of-the-art methods and the outcome of our experiments seems quite conclusive in regard to the merit of the procedure that we propose.

Acknowledgments

This research has been partially supported by the Ministerio de Ciencia e Innov. (Ref. TIN2008-06890-C02-02), and by the Comunidad de Madrid–URJC (Ref. URJC-CM-2008-CET-3731).

References

[1] Baluja, S.: Population-based incremental learning, Carnegie Mellon Univ. Pittsburgh, PA, CMU-CS-94-163 (1994)
[2] Cano, J.R., Herrera, F., Lozano, M.: Using Evolutionary Algorithms as Instance Selection for Data Reduction in KDD: An Experimental Study. IEEE Transactions on Evolutionary Computation~7, 561–575 (2003)
[3] Eshelman, L.J.: The adaptive search algorithm: How to have safe search when engaging in non-traditional genetic recombination. In: Foundations of Genetic Algorithms-1, pp. 265–283. Morgan-Kauffman, San Francisco (1991)
[4] Feo, T.A., Resende, M.: Greedy adaptive search procedures. Journal of Global Optimization~6, 109–133 (1995)
[5] Freitas, A.A.: Data Mining and Knowledge Discovery with Evolutionary Algorithms. Springer, Heidelberg (2002)
[6] Garc\'{\i}a, S., Cano, J.R., Herrera, F.: A memetic algorithm for evolutionary prototype selection: A scaling up approach. Pattern Recognition~41, 2693–2709 (2008)
[7] Glover, F., Laguna, M.: Tabu Search. Kluwer Academic Publishers, Boston (1997)
[8] Han, J., Kamber, M.: Data Mining: Concepts and Techniques. Morgan Kaufmann, San Francisco (2006)
[9] Hansen, P., Mladenov\'{\i}c, N.: Variable Neightborhood Search. In: Glover, F., Kochenberger, G. (eds.) Handbook of Metaheuristics, pp. 145–184. Kluwer, Dordrecht (2003)
[10] Kuncheva, L.I.: Editing for the k-nearest neighbors rule by a genetic algorithm. Pattern Recognition Letters~16, 809–814 (1995)
[11] Liu, H., Motoda, H.: On issues of instance selection. Data Mining and Knowledge Discovery~6, 115–130 (2002)

Expanding Gene-Based PubMed Queries

Sérgio Matos, Joel P. Arrais, and José Luis Oliveira

Abstract. The rapid expansion of the scientific literature is turning the task of finding relevant articles into a demanding one for researchers working in the biomedical field. We investigate the use of a query expansion strategy based on a thesaurus built from standard resources such as the Entrez Gene, UniProt and KEGG databases. Results obtained on the ad-hoc retrieval task of the TREC 2004 Genomics track show that query expansion improves retrieval performance on gene-centered queries. An overall mean average precision of 0.4504 was obtained, which corresponds to an increase of 96% over the use of PubMed as the retrieval engine.

Keywords: Query Expansion, Information Retrieval, Biomedical Literature.

1 Introduction

The rapid expansion of the scientific literature, especially in the biomedical field, is creating many difficulties for researchers, who need to keep informed about their area of work. In fact, since much more information and publications are produced every day, knowing about the latest developments or finding articles satisfying a particular information need is rapidly becoming a demanding task. Structured information, annotated in various biomedical databases, has helped alleviate this problem. However, many relevant research outcomes are still only available in the literature, which remains the major source of information. A significant challenge for researchers is therefore how to identify, within this growing number of publications, the relevant articles for their specific study. This has led to an increasing interest in the application of text mining and information retrieval methods in biomedical literature [1-4].

MEDLINE, the major biomedical literature database, indexes over 17 million citations, which are accessible through the PubMed information retrieval system. PubMed facilitates access to the biomedical literature by combining the Medical

Sérgio Matos · Joel P. Arrais · José Luis Oliveira
Universidade de Aveiro, Campus Universitário de Santiago, 3810-193 Aveiro, Portugal
e-mail: {aleixomatos,jpa,jlo}@ua.pt

M.P. Rocha et al. (Eds.): IWPACBB 2010, AISC 74, pp. 61–68, 2010.
springerlink.com   © Springer-Verlag Berlin Heidelberg 2010

Subject Headings (MeSH) based indexing from MEDLINE, with Boolean and vector space models for document retrieval, offering a single interface from which these articles can be searched [5]. However, and despite these strong points, there are some limitations in using PubMed or other similar tools. A first limitation comes from the fact that user queries are often not fully specified or clear, which is a main problem in any information retrieval (IR) system [6]. This usually means that users will have to perform various iterations and modifications to their queries in order to satisfy their information need. Another drawback is that PubMed does not sort the retrieved documents in terms of how relevant they are for the user query. Instead, the documents satisfying the query are retrieved and presented in reverse date order. This approach is more suitable to users familiar with a particular field who want to find the most recent publications. However, if this is not the case, the most relevant documents may appear too far down the result list to be easily retrieved by the user.

To address the issues mentioned above, several tools have been developed in the past years that combine information extraction, text mining and natural language processing techniques to help retrieve relevant articles from the biomedical literature [4, 7]. Despite the availability of such applications, we feel that the demand for tools that help finding references relevant for a set of genes is still not fully addressed. This constitutes an important query type, as it is a typical outcome of many experimental techniques. The ability to rapidly identify the literature describing these genes and relations among them may be critical for the success of data analysis. In such cases, the problem of obtaining the documents which are more relevant to the user information need becomes even more critical because of the large number of genes being studied, the high degree of synonymy and multiplicity of spelling variations, and the ambiguity in gene names.

This article presents evaluation results of a previously proposed method for expanding gene-based literature queries [8]. The evaluation was performed on a set of 17 pre-selected gene-centered queries from the ad-hoc retrieval task of the TREC 2004 Genomics track. In order to assess the gain obtained with the proposed method, we compare the retrieval performance against the use of PubMed and against searching a local index of MEDLINE, when no query expansion is used.

2 Query Expansion

The query expansion (QE) strategy presented here follows two different perspectives. The first one expands each gene in the query to its known synonyms, in order to deal with the different synonyms and spelling variations that can be found in the literature for a given gene [9, 10]. The second perspective includes gene-related terms or concepts in the query, following the ideas of concept-based query expansion [11, 12]. The approach is based on a thesaurus built from standard resources such as the Entrez Gene, UniProt and KEGG databases, and from well-established relations between the genes and biological concepts in these resources. This is implemented as a local relational database through which a gene symbol can be expanded to all known synonyms and to the related biomedical entities.

Here we present results using gene synonyms, protein names and metabolic pathways as expansion terms, but the approach is general and can be expanded to include other concepts such as diseases or biological processes.

Query expansion works as follows: for each gene symbol in a gene-based query, the alternative gene symbols, protein names and metabolic pathways are obtained from the database. Then, for each of these terms, we search a local MEDLINE index and retrieve the documents matching that term and the respective search score. All results, composed of a document-score pair, are kept on three different lists, one for each class of search terms: gene, protein, and pathway. Finally, all scores for each retrieved document are accumulated, resulting in a final score for that document. The reason for using separate results lists for each concept class is that different weights can be used when accumulating the scores, giving different significance to either class as desired [8]. Fig. 1 illustrates an example of this method.

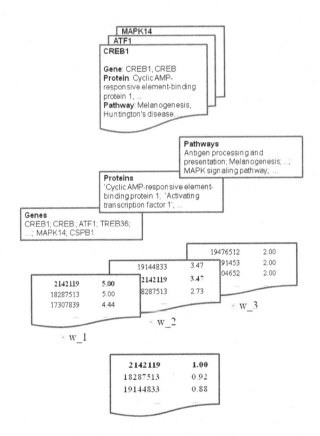

Fig. 1 Query expansion example. Gene query: 'CREB1, ATF1, MAPK14'

3   Results

To evaluate the effectiveness of the proposed method in improving retrieval results for gene-based queries, we tested the performance in terms of the TREC
2004 ad-hoc retrieval task [13]. However, since we follow a gene-centered methodology, general queries such as "Gene expression profiles for kidney in mice" or
"Cause of scleroderma" had to be excluded. From the 50 available queries, 17 that
were found as being more centered in a particular gene (or set of genes) were selected. Furthermore, rather than using a more specified query such as the ones
used in TREC, these queries had to be pre-processed to select just the gene names.
For example, for the queries "Role of TGFB in angiogenesis in skin" and "Role of
p63 and p73 in relation to DNA damage", only the gene name(s) are used for
searching ("TGFB" and "p63, p73", respectively).

As mentioned in the previous section, an important characteristic of the proposed strategy is the use of weights for each class of associated concepts used in
query expansion. Assigning these weights has a considerable effect in the ordering
of retrieved documents and consequently, in retrieval performance. In order to
measure this effect, we empirically selected the best weights for maximizing
performance. We ran different experiments and compared the results to a baseline
obtained by using PubMed as the search engine. In that case, we used the Entrez
e-utils [14] to retrieve a maximum of 10000 documents for each query, limited to
the dates covered in the TREC 2004 document collection (from 1994 to 2003, in-
clusive). The queries used in PubMed were composed of the gene symbols in the
TREC queries, joined by a disjunction ('OR') in the cases of queries with more
than one gene. The results obtained are shown in Fig. 2 and Fig. 3. For the other
results shown, we used a local index of the MEDLINE database, created with the
Lucene indexing software [15]. We calculated the retrieval performance using the
genes in the queries (no expansion), using just the gene synonyms in the expansion, using all concepts, and finally, using all concepts and empirically selecting,
for each individual query, the weights that maximize the mean average precision
(MAP) value.

Using the QE methodology proposed, a mean average precision of 0.4504 can
be achieved for these 17 queries, when the best weights, for each concept class,
are selected for each query. This compares to a MAP value of 0.2295 when using
PubMed (an increase of 96%) and to 0.2920 when the local MEDLINE index is
searched without any query expansion (an increase of 54%). Using gene synonyms in the query already improved the results to a MAP value of 0.3276, corresponding to a 12% increase as compared to no expansion, and 43% as compared
to PubMed.

Also shown are other metrics commonly used for assessing retrieval performance: reciprocal rank, which measures the inverse rank of the first relevant article
retrieved (average 0.6526 versus 0.1961 in PubMed and 0.7005 for no expansion);
precision after ten documents retrieved (0.5118 versus 0.1412 and 0.3882); precision after 100 documents retrieved (0.3412 versus 0.1659 and 0.2365); and recall
after 1000 documents (0.6982 versus 0.5521 and 0.4426). The reciprocal rank

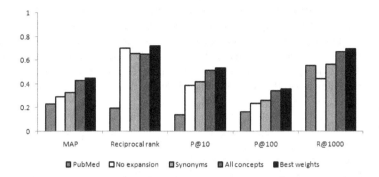

Fig. 2 Performance results on the TREC 2004 ad-hoc retrieval task: mean average precision (MAP); reciprocal rank; precision after 10 and 100 documents; and recall after 1000 documents

results show that, in average in these 17 queries, only the fifth document in the PubMed results is a relevant document, according to the TREC data. In comparison, in our results, either the first of second documents retrieved are, in average, relevant. This difference is related to the fact that the results returned by PubMed are not ranked by relevance. Another interesting result is that the reciprocal rank obtained from searching the index with no query expansion, i.e. using just the gene symbol, is slightly higher than when synonyms or associated concepts are added to the query (Fig. 2). This reflects the fact that QE may lead to query drift, which leads to non-relevant documents being added to the results. On the other hand, we can also see from the results that the weighting mechanism allows alleviating this problem and improves the results.

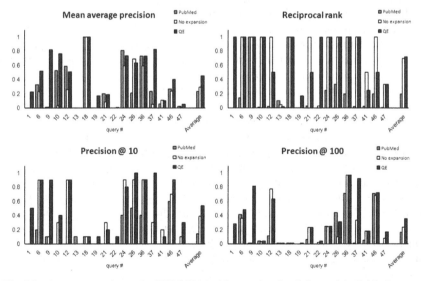

Fig. 3 Performance results on the TREC 2004 ad-hoc retrieval task for each individual query

As can be seen from the results, although searching the local index using just the gene symbol(s) in the query gives an improvement in all other statistics, as compared to PubMed, the obtained recall is significantly lower. However, when gene synonym expansion was used, the recall achieved the same level as with PubMed (0.5620). The best recall was obtained with the proposed QE method, with an improvement of 26% over the use of PubMed.

4 Discussion

We present the evaluation results of a query expansion method for gene-based literature queries. The proposed approach tries to address many problems related to genomic information retrieval, namely: how to deal with the various names and symbols used for a particular gene [9-10, 16]; how to address the user information need in terms of what are the concepts related to the genes that the user is most interested in; and what is the best query expansion strategy for this specific problem, particularly, what are the best terms to use for expanding the query and how should these terms contribute to the expanded query results [16, 17].

The results presented here indicate good overall performance of the proposed method, especially when compared to the use of PubMed. Although PubMed uses query expansion through Automatic Term Mapping (ATM), and includes manually annotated information in the search, through MeSH terms [18], our evaluation results show a 96% increase in mean average precision (0.4504) as compared to PubMed (0.2295). The reduced performance obtained with PubMed in this type of task, and using these performance measures, is related to the lack of any relevance ranking of the resulting documents, leading to relevant documents appearing lower in the returned list. However, since PubMed is still the most popular literature retrieval tool used by biomedical researchers, we feel that using it as a baseline for comparing our results is a valid approach. Also, evaluation metrics such as the mean average precision try to reflect how effectively the retrieval results satisfy a particular information need expressed in the query, giving more emphasis to relevant documents that appear higher in the results list. Additionally, our results also show significant improvements in terms of precision and recall.

Although performance measures based on a manually annotated test set give some indication of the method's retrieval performance, the 2004 TREC genomics evaluation data and methodology may not be entirely suited for testing our method. First of all, it is oriented to a very specific query type as are gene-based queries. Additionally, not all query types in the TREC task could be used. Since this is a gene-centered method, we had to exclude the majority of the queries, ending up with 17 which were considered to be more focused on a particular gene (or set of genes). These limitations imply that the results obtained do not allow a completely valid comparison with other systems that use the same evaluation, including the ones that participated in the TREC task and other more recent ones. Nonetheless, we consider that these results and the analysis presented here show that this method can be used to improve retrieval results for this particular type of queries.

A common problem in genomic information retrieval is that of ambiguity, that is, the same symbol identifying different genes and/or proteins. The query expansion method discussed here helps addressing this problem through the inclusion of related terms in the query. This is because documents containing more terms associated to the input genes will have a higher ranking. For example, an abstract containing the terms 'CAT' and 'catalase' will have higher relevance than an abstract containing just the ambiguous term 'CAT'. We plan to further explore this aspect by introducing reliability (or ambiguity) scores to the terms in our thesaurus.

5 Conclusions

This paper presents an evaluation of a concept-oriented query expansion methodology for searching the MEDLINE literature database using gene-based queries. The approach allows finding documents containing concepts related to the genes in the user input, such as proteins and pathway names. Results obtained using the TREC 2004 ad-hoc retrieval data show a considerable improvement over the use of PubMed, as measured by the MAP statistic.

Acknowledgments. The research leading to these results has received funding from the European Community's Seventh Framework Programme (FP7/2007-2013) under grant agreement n° 215847 - the EU-ADR project. S. Matos is funded by Fundação para a Ciência e Tecnologia (FCT) under the Ciência2007 programme.

References

1. Altman, R., Bergman, C., Blake, J., Blaschke, C., Cohen, A., Gannon, F., Grivell, L., Hahn, U., Hersh, W., Hirschman, L., et al.: Text mining for biology - the way forward: opinions from leading scientists. Genome Biol. 9(Suppl. 2), S7 (2008)
2. Jensen, L.J., Saric, J., Bork, P.: Literature mining for the biologist: from information retrieval to biological discovery. Nat. Rev. Genet. 7, 119–129 (2006)
3. Rebholz-Schuhmann, D., Kirsch, H., Couto, F.: Facts from text–is text mining ready to deliver? PLoS Biol. 3(2), e65 (2005)
4. Krallinger, M., Valencia, A., Hirschman, L.: Linking genes to literature: text mining, information extraction, and retrieval applications for biology. Genome Biol. 9(Suppl.2), 8 (2008)
5. PubMed, http://www.ncbi.nlm.nih.gov/pubmed
6. Manning, C., Raghavan, P., Schütze, H.: Introduction to Information Retrieval. Cambridge University Press, New York (2008)
7. Weeber, M., Kors, J.A., Mons, B.: Online tools to support literature-based discovery in the life sciences. Brief Bioinform. 6(3), 277–286 (2005)
8. Arrais, J., Rodrigues, J., Oliveira, J.: Improving Literature Searches in Gene Expression Studies. In: Proceedings of the 2nd International Workshop on Practical Applications of Computational Biology and Bioinformatics (2009)
9. Chen, L., Liu, H., Friedman, C.: Gene name ambiguity of eukaryotic nomenclatures. Bioinformatics 21, 248–256 (2005)

10. Koike, A., Takagi, T.: Gene/protein/family name recognition in biomedical literature. In: Proceedings of BioLINK 2004: linking biological literature, ontologies, and databases. Association for Computational Linguistics (2004)
11. Lu, Y., Fang, H., Zhai, C.: An empirical study of gene synonym query expansion in biomedical information retrieval. Inf. Retr. 12, 51–68 (2009)
12. Qiu, Y., Frei, H.-P.: Concept based query expansion. In: Proceedings of the 16th annual international ACM SIGIR conference on Research and development in information retrieval. ACM, Pittsburgh (1993)
13. Hersh, W.R., Bhupatiraju, R.T., Ross, L., Roberts, P., Cohen, A.M., Kraemer, D.F.: Enhancing access to the Bibliome: the TREC 2004 Genomics Track. J. Biomed. Discov. Collab. 13, 1–3 (2006)
14. Entrez Programming Utilities,
http://eutils.ncbi.nlm.nih.gov/corehtml/query/static/
eutils_help.html
15. Apache Lucene, http://lucene.apache.org/
16. Lu, Y., Fang, H., Zhai, C.: An empirical study of gene synonym query expansion in biomedical information retrieval. Inf. Retr. 12(1), 51–68 (2009)
17. Stokes, N., Li, Y., Cavedon, L., Zobel, J.: Exploring criteria for successful query expansion in the genomic domain. Inf. Retr. 12(1), 17–50 (2009)
18. Lu, Z., Kim, W., Wilbur, W.J.: Evaluation of Query Expansion Using MeSH in PubMed. Inf. Retr. 12(1), 69–80 (2009)

Improving Cross Mapping in Biomedical Databases

Joel Arrais, João E. Pereira, Pedro Lopes, Sérgio Matos, and José Luis Oliveira

Summary. The complete analysis of many large scale experiments requires the comparison of the produced output with data available in public databases. Because each datapbase uses its own nomenclature to classify entries, this task frequently implies the conversion of identifiers and, due to incomplete mapping between those identifiers, this tasks commonly causes loss of information.

In this paper, we propose a methodology to improve the coverage of the mapping between database identifiers. As a starting point we use a local warehouse with the default mappings from the most relevant biological databases. Next we apply a methodology to four database identifiers (Ensembl, Entrez Gene, KEGG and UniProt). The results showed an improvement in the coverage of all relationships superior to 10% in three and to 7% in five relations.

Keywords: Biomedical databases, identifiers mapping.

1 Introduction

The integration of heterogeneous data sources has been a fundamental problem in database research over the last two decades [1]. The goal is to achieve better methods to combine data residing at different sources, under different schemas and with different formats in order to provide the user with a unified view of the data. Although simple in principle, several constrains turn this into a very challenging task where both the academic and the commercial communities have been working and proposing several solutions that span a wide range of fields.

Life sciences is just one of many fields that take advantage from the advances in data integration methods [2]. This is because the information that describes genes, gene products and the biological processes in which they are involved are dispersed over several databases [3]. In addition, due to the advances in some high throughput techniques, such as microarrays, the experimental results obtained in the laboratory are only valuable after being matched with data stored in public d atabases [4].

Joel Arrais · João E. Pereira · Pedro Lopes · Sérgio Matos · José Luis Oliveira
University of Aveiro, DETI/IEETA, 3810-193 Aveiro, Portugal
e-mail: {jpa, jlo}@ua.pt

M.P. Rocha et al. (Eds.): IWPACBB 2010, AISC 74, pp. 69–76, 2010.
springerlink.com   © Springer-Verlag Berlin Heidelberg 2010

One major issue when combining data from different sources consists in establishing a match between the identifiers for the same biological entities. This happens because each database uses its own nomenclature to classify entries, resulting
in a multitude of identifiers for the same entity. For instance, the gene BRCA2
has, among others, the following identifiers: hsa:675 for the KEGG [5] database;
675 for the Entrez Gene [6] database; ENSG00000139618 for the Ensembl [7]
database; and P51587 for the associated protein in UniProt [8] database.

In spite of the recent notorious effort from major databases to establish associations between their identifiers, the resulting cross-database mapping is still far
from perfect due to the low number of matches and included databases. This low
or absent coverage creates major obstacles as some biological entities are discarded in cross-mappings and, as such, possible meaningful biological results may
be overlooked. As a possible example, consider the analysis of the distribution of
a set of Ensembl genes to KEGG Pathways. Since such an analysis would require
an initial mapping from Ensembl to KEGG identifiers, it is possible to miss some
genes and therefore, pathway associations.

In this paper, we propose a simple yet effective methodology to improve the
coverage of cross-database mapping. We applied this approach to the identifiers of
the four most relevant databases: Ensembl, Entrez Gene, KEGG and UniProt. As a
back-end resource we use GeNS (Genomic Name Server) [9] which consists of a
extendable platform for the storage of biological identifiers.

2 Related Work

According to the last release of the Nucleic Acids Research *Molecular Biology
Database Collection* there are about 1000 databases in the field of molecular biology [3]. Each database corresponds to the output of a specific study or community
and represents a huge investment whose potential has not been fully explored.

Being able to integrate multiple sources is important because data about one
biological entity may be dispersed over several databases. For instance, for a gene,
the nucleotide sequence is stored in GenBank , the protein structure on PDB (Protein DataBank), the pathway in KEGG Pathway and the expression data in
ArrayExpress. Obtaining a unified view of these data is therefore crucial to
understanding the role of the gene.

Much of the work necessary to obtain this "unified view" consists in establishing relationships between the identifiers of each database. This identifiers mapping is already provided by several public databases including KEGG LinkDB and
the UniProt ID Mapping, however, they lack global completeness and reliability.
One example of such is KEGG and Ensembl that provide gene information and
are supposed to cross-reference each other. Other problem consists in the necessity
to navigate across several databases to find a relationship between two databases.
Those problems are illustrated in Figure 1. For each identifier a graph can be built
with the existing relationships in each specific database. For instance, in the database Ensembl the gene "keratin associated protein 5-6" with the identifier
ENSG00000205864 does not have a direct link to the database KEGG, despite the

fact that KEGG has a link to Ensembl. The link from Ensembl to KEGG can also be inferred with the intermediate databases HGNC and UniProt.

To address this issue, several efforts to establish unified maps of the various databases have been proposed [10-14] including Onto-Translate [15], IDconverter [16] and MatchMiner [17].

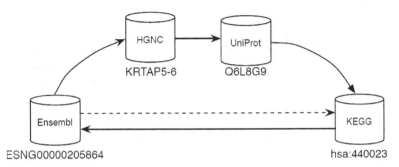

Fig. 1 Example of the lack of completeness in the network of associations between databases

Onto-Translate integrates 17 annotation databases and enables many-to-many conversion between stored identifiers. Given a source identifier and a target identifier type, the implemented algorithm obtains a target identifier based on a best match determined via the database trustworthiness. Despite the comprehensive list of included identifiers, it does not include Ensembl, Pubmed and Reactome Pathways identifiers, for example.

IDconverter is another tool that, with a simple query, allows mapping multiple identifiers into multiple outputs. Although the list of connected databases is broad, major limitations are found in the input list because it only accepts gene/protein identifiers. Therefore, if the user has a list of OMIM or pathway identifiers this tool cannot be used. Another limitation is its low number of covered species (Human, Mouse and Rat).

Similarly to IDconverter, the MatchMiner annotation tool is restricted to genes as input identifiers. Moreover, this tool's output is even more restricted as it does not include relationships to the Gene Ontology, KEGG and PFAM databases, to name just a few. It is also limited to the Human species.

3   Implementation

Based on the previous considerations, we propose a methodology to improve the issue of cross-database mapping. As a starting point, we use the GeNS [9], a local database that contains the most relevant biological resources, including UniProt, Entrez Gene, KEGG, Ensembl and Gene Ontology (Figure 2). By merging these data, we have obtained approximately one thousand species, with over 7 million gene products, 70 million alternative gene/protein identifiers and 70 million associations to 140 distinct biological entities. For instance, the species *Saccharomyces*

Cerevisiae has 7421 gene products with 105.000 alternative identifiers and associations with 213.000 biological entities such as pathways, Gene Ontology terms or homologs.

GeNS presents a set of characteristics that turn it a good case study for our goal: firstly, despite being focused on storing biological identifiers, it also contains data commonly stored in warehouses such as BioWarehouse [18]; secondly, it integrates a large number of databases where the lack of full cross-reference is evident, and thirdly, it has a generic database schema that allows effortless addition of new databases and relations (hence, without the need to change the current schema). This generic schema also allows the implementation of a generic interface for searching, retrieving and inserting elements, thus enabling easy implementation of recursive algorithms such as the one used in the proposed methodology.

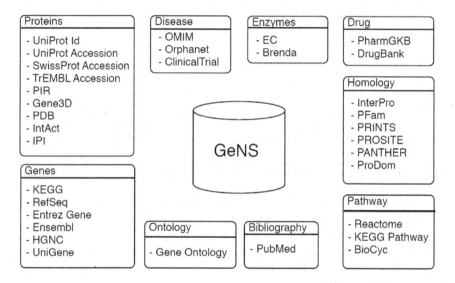

Fig. 2 Biological entities and identifiers stored in GeNS database

Considering each database as a node and each database reference as a connection, we have created, for each entry, a digraph that represents real linkage across databases. To be able to map between the identifier ID1 in database DB1 to ID2 in DB2, one needs to know if there is a direct connection between them. The algorithm acts as follows:

- Firstly, it verifies the existence of a direct connection from (ID1, DB1) to (ID2, DB2). If it exists, the algorithm terminates successfully; if not, it proceeds to the second step;
- Secondly, the algorithm will test the existence of a connection from (ID2, DB2) to (ID1, DB1). If this connection exists, the algorithm will create a reverse connection and terminate successfully; if not, it will proceed to step three;

- Thirdly, it will repeat the procedure iterating through the nodes directly connected with DB1. When a node that connects to (ID2, DB2) is found, the algorithm adds a new direct connection from (ID1, DB1) to (ID2, DB2) and terminates.

The algorithm's pseudocode is detailed next – where *node* is a pair (identifier, database). It is self-explanatory and its inner functions should implement all the non-described operations.

```
Function mapping(node1, node2)
    If node1.hasConnectionTo(node2)
       Return true

    If node2.hasConnectionTo(node1)
       node1.createConnection(node2)
       Return true

    For each node in node1.connectionList
       If mapping(node, node2) then
          node1.createConnection(node2)
          Return true

    Return false
```

4 Analysis of the Coverage Improvements

To analyse the presented methodology we have selected the identifiers from four databases: UniProt Acession, KEGG, Ensembl and Entrez Gene. Altogether, these four databases store links to more than one hundred distinct external databases and, therefore, by improving the coverage between these four identifiers, one is able to extend his search to all the associated ones. We have also restricted this analysis to one organism (*Homo sapiens*) in order to make this example as simple as possible.

The initial step consisted of measuring the coverage value of each database. By this we mean the percentage of entries in the origin database that have direct correspondence in the target database. Table 1 shows the coverage results, for any possible connection, before and after executing the algorithm. Overall, a significant improvement in the coverage has been obtained in every relationship; Entrez Gene data showed the greater increase essentially due to its low initial coverage: the links from Entrez Gene to UniProt noticed a positive difference of 10,4%, while those to Ensembl increased by 10,2% and to KEGG by 8,0%. Other relevant improvements were registered in the Ensembl to KEGG relationship (38,9%) and in the UniProt to Entrez Gene relationship (7,4%).

We have also compared the results with three tools for database mapping: Onto-Translate, IDConverter and MatchMiner. As input list we used Entrez Gene identifiers annotations for the Affymetrix GeneChip® Human Genome U133 Plus 2.0 Array (19 151 unique identifiers). The graph in Figure 3 shows the coverage of each tool, for each translation. We notice that GeNS obtained an overall high coverage with a minimum value of 89,9% for the Ensembl identifiers. However, this value is still higher than the one from IDConverter (81,9%), while Onto-Translate and MatchMiner do not allow translating to Ensembl identifers. The major difference found on IDConverter while translating to UniProt (69,7%) is because it only maps to SwissProt. Apart from that, all tools present very high and uniform results.

Table 1 Comparison of the average coverage values with and without the use of the cross-reference algorithm. For each relationship between two databases the coverage values before and after the algorithm are shown

TO FROM	UniProt		KEGG		Ensembl		Entrez Gene	
	Before	After	Before	After	Before	After	Before	After
UniProt	–	–	83,8%	83,9%	97,1%	97,4%	85,7%	**93,1%**
KEGG	76,8%	**80,5%**	–	–	77,2%	79,8%	100%	100%
Ensembl	97,7%	97,9%	34,3%	**73,2%**	–	–	88,9%	89,2%
Entrez Gene	47%	**57,4%**	56,3%	**64,3%**	46,8%	**57%**	–	–

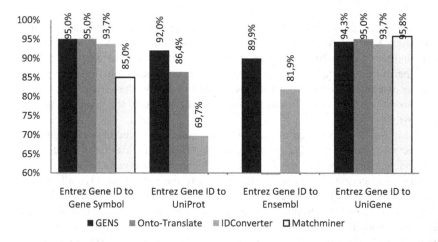

Fig. 3 Coverage comparison for the GENS, Onto-Translate, IDConverter and MatchMiner. The input file consists of 19 151 genes (Entrez gene identifiers) obtained from the Affymetrix GeneChip® Human Genome U133 Plus 2.0 Array. The graph shows the percentage of genes with at least one match for each output format. The empty bars correspond to translations not provided

5 Usage Example

Taking into consideration the previous mentioned example, a user is trying to determine the KEGG identifier for the human protein *Keratin-associated protein 5-6* using the Ensembl identifier ENSG00000205864 as a starting point. Before running the algorithm on the database, no direct match could have been established between these two identifiers; nevertheless, we can now obtain matches between Ensembl and KEGG identifiers. This was achieved because, as seen in Figure 1, the HGNC and UniProt accession Q6L8G9 can be used to establish a direct link between the two.

GeNS database provides a set of public web services that allow external applications and users to better exploit its concepts, associations and sources[1]. One of these methods, *ConvertIdentifier*, enables real-time identifier conversion and can be used to retrieve identifier associations. As such, the previous example can be replicated through this API using a common web browser. In this case one must run *ConvertIdentifier* with the parameters *(9606, KRTAP5-6, 27, 9)*, where "9606" refers to the taxonomic identifier of the *Homo sapiens* species, "27" to an internal data-type identifier that states the source of that input as belonging to **Ensembl** and "9" to specify the output format of the identifier as a KEGG ID. These IDs can be easily obtained from other web services. Finally, GeNS would then return "hsa:440023" in XML format, thus bridging the gap between these distinct identifiers in the original databases.

6 Conclusion

In this paper, we have presented a methodology to improve the coverage of cross database mapping. Such a methodology is relevant due to the low coverage between certain types of identifiers which, in turn, can result in some biological entities being discarded in cross-mappings, thus possibly ignoring meaningful biological results. This methodology has been implemented over the GeNS database that already stores identifiers for the most relevant biological databases in order to further increase its coverage.

We have shown the gain of the methodology with two distinct analyses. In the first one, we compare the improvement to the coverage with the default links provided by the used databases; we have attained an improvement in all relations being superior to 10% in three and to 7% in five relations. This is relevant because eight of the initial values were already superior to 75%. In the second analysis, we compare the database mapping against three of the most used mapping tools with a gene list from the Affymetrix U133 Plus 2.0 array. In this case we also attained significant improvements, especially when converting to UniProt and Ensembl identifiers.

---

[1] http://bioinformatics.ua.pt/applications/gens/

Acknowledgement

The research leading to these results has received funding from the European Community's Seventh Framework Programme (FP7/2007-2013) under grant agreement n° 200754 - the GEN2PHEN project. S. Matos is funded by Fundação para a Ciência e Tecnologia (FCT) under the Ciência2007 programme.

References

1. Widom, J.: Research problems in data warehousing. In: Proceedings of the fourth international conference on Information and knowledge management, pp. 25–30. ACM, Baltimore (1995)
2. Goble, C., Stevens, R.: State of the nation in data integration for bioinformatics. J. Biomed. Inform. 41(5), 687–693 (2008)
3. Galperin, M.Y.: The Molecular Biology Database Collection: 2008 update. Nucleic Acids Res. (2007)
4. Al-Shahrour, F., et al.: From genes to functional classes in the study of biological systems. BMC Bioinformatics 8, 114 (2007)
5. Kanehisa, M., et al.: KEGG for linking genomes to life and the environment. Nucleic Acids Res. (2007)
6. Maglott, D., et al.: Entrez Gene: gene-centered information at NCBI. Nucleic Acids Res. 35(Database issue), D26–D31 (2007)
7. Flicek, P., et al.: Ensembl 2008. Nucleic Acids Res. 36(Database issue), 707–714 (2008)
8. Wu, C.H., et al.: The Universal Protein Resource (UniProt): an expanding universe of protein information. Nucleic Acids Res. 34(Database issue), D187–D91 (2006)
9. Arrais, J., et al.: GeNS: a biological data integration platform. In: International Conference on Bioinformatics and Biomedicine, venice, Italy (2009)
10. Diehn, M., et al.: Source: a unified genomic resource of functional annotations, ontologies, and gene expression data. Nucleic Acids Res. 31(1), 219–223 (2003)
11. Tsai, J., et al.: Resourcerer: a database for annotating and linking microarray resources within and across species. Genome Biol. 2(11) (2001) Software0002
12. Lenhard, B., Wahlestedt, C., Wasserman, W.W.: GeneLynx mouse: integrated portal to the mouse genome. Genome Res. 13(6B), 1501–1504 (2003)
13. Castillo-Davis, C.I., Hartl, D.L.: GeneMerge–post-genomic analysis, data mining, and hypothesis testing. Bioinformatics 19(7), 891–892 (2003)
14. Zhang, J., Carey, V., Gentleman, R.: An extensible application for assembling annotation for genomic data. Bioinformatics 19(1), 155–156 (2003)
15. Draghici, S., et al.: Onto-Tools, the toolkit of the modern biologist: Onto-Express, Onto-Compare, Onto-Design and Onto-Translate. Nucleic Acids Res. 31(13), 3775–3781 (2003)
16. Alibes, A., et al.: IDconverter and IDClight: conversion and annotation of gene and protein IDs. BMC Bioinformatics 8, 9 (2007)
17. Bussey, K.J., et al.: MatchMiner: a tool for batch navigation among gene and gene product identifiers. Genome Biol. 4(4), R27 (2003)
18. Lee, T.J., et al.: BioWarehouse: a bioinformatics database warehouse toolkit. BMC Bioinformatics 7, 170 (2006)

An Efficient Multi-class Support Vector Machine Classifier for Protein Fold Recognition

Wiesław Chmielnicki, Katarzyna Stąpor, and Irena Roterman-Konieczna

Abstract. Predicting the three-dimensional (3D) structure of a protein is a key problem in molecular biology. It is also interesting issue for statistical methods recognition. In this paper a multi-class Support Vector Machine (SVM) classifier is used on a real world data set. The SVM is a binary classifier and how to effectively extend a binary to the multi-class classifier case is still an on-going research problem. The new efficient approach is proposed in this paper. The obtained results are promising and reveal areas for possible further work.

1 Introduction

Predicting the three-dimensional (3D) structure of a protein is a key problem in molecular biology. Proteins manifest their function through these structures so it is very important to know not only sequence of amino acids in a protein molecule, but also how this sequence is folded.

There are several machine-learning methods to detect the protein folds from amino acids sequences proposed in literature. Ding and Dubchak [5] experiments with Support Vector Machine (SVM) and Neural Network (NN). Shen and Chou [17] proposed ensemble model based on nearest neighbour. A modified nearest neighbour algorithm called K-local hyperplane (HKNN) was used by Okun [16]. Nanni [15] proposed Ensemble of classifiers: Fishers linear classifier and HKNN classifier.

Wiesław Chmielnicki
Faculty of Physics, Astronomy and Applied Computer Science
e-mail: wieslaw.chmielnicki@uj.edu.pl

Katarzyna Stąpor
Silesian University, Institute of Computer Science

Irena Roterman-Konieczna
Jagiellonian University, Faculty of Medicine,
Department of Bioinformatics and Telemedicine

M.P. Rocha et al. (Eds.): IWPACBB 2010, AISC 74, pp. 77–84, 2010.
springerlink.com

This paper focuses on the SVM classifiers. The SVM technique has been used in different application domains and has outperformed the traditional techniques in erms of generalization capability. However, the SVM is a binary classifier but the protein fold recognition is a multi-class problem and how to effectively extend a binary to the multi-class classifier case is still an on-going research problem. There are many methods proposed to deal with this issue. The most popular is to construct a multi-class classifier by combining binary classifiers.

The number of classes in protein fold problem can be as high as 1000, so some of these strategies might not be applicable according to the number of two-way classifiers needed. In this paper the new strategy is presented, which minimizes the number of two-way classifiers. The results using this method are promising especially for the problems with large number of classes.

The rest of this paper is organized as follows: Section 2 introduces the database and the feature vectors used is these experiments, Section 3 shortly describes basics of SVM classifier, Section 4 deals with different ways of solving of multi-class problem, Section 5 describes the proposed approach to multi-class classifier and Section 6 present experimental results and conclusions.

2   The Database and the Feature Vectors

In experiments described in this paper two data sets derived from SCOP (Structural Classification of Proteins) database are used. The detailed description of these sets can be found in Ding and Dubchak [5]. The training set consists of 313 protein sequences and the testing set consists of 385 protein sequences. These data sets include proteins from 27 most populated different classes (protein folds) representing all major structural classes: α, β, α/β and $\alpha + \beta$. The training set was based on PDB_select sets (Hobohm and Sander [9]) where two proteins have no more than 35% of the sequence identity. The testing set was based on PDB-40D set developed by Lo Conte et al. [13] from which representatives of the same 27 largest folds are selected. The proteins that had higher than 35% identity with the proteins of the training set are removed from the testing set.

In this paper the feature vectors developed by Ding and Dubchak [5] were used. These feature vectors are based on six parameters: Amino acids composition (C), predicted secondary structure (S), Hydrophobity (H), Normalized van der Waals volume (V), Polarity (P) and Polarizability (Z). Each parameter corresponds to 21 features except Amino acids composition (C), which corresponds to 20 features. The data sets including these feature vectors are available at http://ranger. uta.edu/~chqding/protein/. For more concrete details, see Dubchak et al. [6]. In this study the sequence length was added to the Amino acids composition (C) vector and the feature vectors based on these parameters were used in different combinations creating vectors from 21D to 126D.

3 The SVM Classifier

The Support Vector Machine (SVM) is a well known large margin classifier proposed by Vapnik [18]. The basic concept behind the SVM classifier is to search an optimal separating hyperplane, which separates two classes. The perfect separation is not often feasible, so slack variables ξ_i can be used which measure the degree of misclassification. Let us consider a classifier whose decision function is given by:

$$f(x) = sign(x^T w + b) , \tag{1}$$

where x denotes a feature vector and w is a weight vector. Then the SVM algorithm minimizes the objective function:

$$\frac{1}{2}\|w\|_2^2 + C \sum_{i=1}^{n} \xi_i , \tag{2}$$

subject to: $y_i(wx_i + b) \geq 1 - \xi_i, \xi_i > 0, i = 1, 2, \ldots, n$.

This problem leads to so called dual optimization problem and finally (considering non-linear decision hyperplane and using the kernel trick) to:

$$f(x) = sign\left(\sum_{i=1}^{N} \alpha_i y_i K(x_i, x) + b \right) , \tag{3}$$

where $0 \leq \alpha_i \leq C, i = 1, 2, \ldots, N$ are nonnegative Lagrange multipliers, C is a cost parameter, that controls the trade off between allowing training errors and forcing rigid margins, x_i are the support vectors and $K(x_i, x)$ is the kernel function.

4 A Multi-class SVM

The SVM is a binary classifier, but protein fold recognition is a multi-class problem. Generally, there are two types of approaches to this problem but one of them is considering all classes in one optimization (Lee et al. [12]). In such a case a QP (quadratic problem) with $(n-1)k$ variables, where n is the number of classes, must be solved, while using binary SVM there is the QP with k variables. Wang and Shen [20] proposed a L1-norm MSVM (Multi category SVM), which performs classification and feature selection simultaneously. This algorithm is more effective but still when the number of classes increases this approach can be computationally very expensive.

The second approach is to cover one n-class problem into several binary problems. There are many methods proposed in literature, such as one-versus-others, one-versus-one strategies, DAG (Directed Acyclic Graph), ADAG (Adaptive Directed Acyclic Graph) methods (Kijsirikul et al. [11]), BDT (Binary Decision Tree) approach (Fei and Liu [7]), DB2 method (Vural and Dy [19]), pairwise coupling (Hasti and Tibshirani [8]) or error-correcting output codes (Dietterich and Bakiri [4]).

4.1   One-versus-One Method

One of the first and well-known methods is one-versus-one method with max-win
voting strategy. In this method the two-way classifiers are trained between all possible pairs of classes and there are $N*(N-1)/2$ of them. All proteins are tested
against these classifiers and then each classifier votes for a preferred class. The
protein with maximum number of votes is classified as the correct class.

The number of two-way classifiers in this method is $N*(N-1)/2$. In these
experiments there are 27 classes used, so the number of two-way classifiers is
$27*(27-1)/2 = 351$. However, the total number of protein folds is estimated as
many as 1000, so the number of classifiers would be $1000*(1000-1)/2 = 499500$.
The recognition process using about half a million two-way classifiers would be
entirely inefficient.

4.2   Binary Decision Tree Method

This method limits the number of two-way classifiers, so it is more suitable for
problems in which the number of classes is large. The classifiers are arranged in a
binary tree structure, each SVM in each node is trained using two sets of classes
and in the root node all classes are divided into two sets. The number of classes in
each set is equal (when the total number of classes is even) or almost equal (when
the total number or classes is odd). Then this procedure is repeated recursively until
all sets contain one class only.

The number of classifiers used in this architecture is $N-1$, however these $N-1$
classifiers are needed to be trained, but at most $\lceil log_2N \rceil$ are required to classify a
sample. When the 1000 class problem is considered there must be 999 classifiers
trained but only 10 are used to classify each sample, compared with the 499500 in
one-versus-one method.

The main problem in this method is how to divide classes into subsets and there
are several approaches in literature. For example Madzarov et al. [14] propose a
kind of similarity measure based on gravity centres of the classes. Jinbai et al. [10]
divide randomly the classes, but they used a strategy for error correcting after each
classification.

A similar strategy was presented by Vural and Dy [19]. This method, called DB2
(Divide-by-2), hierarchically divides the data into two subsets until every subset
consists of one class only. They considered the division step as a clustering problem
and proposed three methods of solving it. One method is to use k-means clustering,
the second is to use class mean distances from origin and the third is to use balanced
subsets (such subsets that the difference of the samples in each subset is minimum).

5   The Modified Approach to Multi-class Classifier

Let us consider two binary classifiers. For example: one dividing classes into $1,2,4$
and $3,5,6,7$ subsets and second dividing classes into $2,4,5$ and $1,3,6,7$ subsets.

It is seen that the intersection of these two classifiers can be used to obtain four subsets: $2, 4 - 3, 6, 7 - 1 - 5$. See fig. 1 (left). This procedure can be recursively used for subsets containing more than one class until all classes are recognized. See fig. 1 (right). The number of classifiers used in this architecture is $N - 1$ in training stage but at most $\lceil log_2 N \rceil$ are required to classify a sample. However some samples are classified using only two binary classifiers (classes 1 and 5 in our example).

Another approach is to continue the procedure, dividing all classes into two different subsets. We assume that both subsets must consist of equal number of classes (or almost equal if the total number of classes is odd). After using two more binary classifiers all classes can be recognized using described intersection scheme. See fig. 2. The number of two-way classifiers used in this strategy is (N + 1) / 2, so it is very useful for applications with a large number of classes. The difference between this strategy and recursive strategy is visible when the number of binary classifiers to recognize all samples is considered. Using recursive strategy $N - 1$ binary classifiers are needed compared to $(N + 1)/2$ when half-versus-half strategy is used.

Unfortunately the recognition rates obtained using described methods are not satisfactory. They are even lower than using one-versus-one method. However, they can be enhanced using different partitions (and corresponding different binary classifiers) with the voting scheme. The multi-class classifier using a set of partitions votes for the preferred classes and then the next multi-class classifier using a different set of partitions is used. Finally the samples with maximum number of votes are assigned to the correct classes.

The results of using this strategy are presented in table 1. There are presented results obtained using the classifier with the recursive procedure and the classifier with half-versus-half strategy. Every classifier was tested without the voting scheme and with the voting scheme using 20, 40 and 60 randomly chosen partitions.

Using different sets of partitions and the voting scheme the recognition rate can be improved but however at the cost of efficiency. Despite this drawback the proposed method allows to control the number of binary classifiers needed to classify the samples.

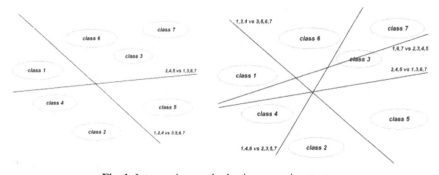

Fig. 1 Intersection method using recursive strategy

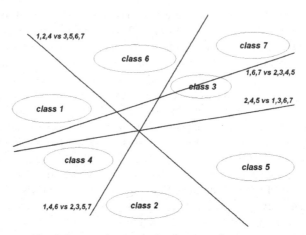

Fig. 2 Intersection method using recursive strategy

Table 1 Results using proposed method

	No voting scheme	Voting scheme using 20 sets	Voting scheme using 40 sets	Voting scheme using 60 sets
Recognition rate using recursive procedure	48.1%	53.5%	56.1%	57.1%
Recognition rate using half-versus-half strategy	50.1%	53.2%	56.4%	57.7%

The well-known LIBSVM library version 2.89 was used in our research (Chang and Lin [2]). Although the implementation of this library includes one-versus-one strategy for the multi category problems only the binary version of the classifier was used. LIBSVM provides a choice of build-in kernels i.e. Linear, Polynomial. Radial Basis Function (RBF) and Gausian. The RBF kernel:

$$K(x_i, x) = -\gamma \|x - x_i\|^2, \gamma > 0 , \tag{4}$$

gave the best results in our experiments.

The parameters C and γ must be chosen to use the SVM classifier with RBF kernel. It is not known beforehand which C and γ are the best for one problem. Both values must be experimentally chosen, which was done by using cross-validation procedure on the training data set. The best recognition ratio was achieved using parameters values $\gamma = 0.1$ and $C = 128$.

6 Results

In this paper there is a new method presented to deal with the multi-class SVM. The proposed classifier was used to solve a protein fold recognition problem. This approach minimizes the number of binary classifiers used, so it is very useful with problems where the total number of classes is big. The achieved recognition ratio is worse than the result achieved using one-versus-one method but however there is a strategy proposed to increase recognition efficiency. The best result (57.7%) is achieved using intersection method with half-versus-half strategy and voting scheme using 60 different subsets.

The accuracy measure used in this paper is the standard Q percentage accuracy (Baldi et al., [1]). Suppose there is $N = n_1 + n_2 + \ldots + n_p$ test proteins, where n_i is the number of proteins which belongs to the class i. Suppose that c_i of proteins from n_i are correctly recognised (as belonging to the class i). So the total number of $C = c_1 + c_2 + \ldots + c_p$ proteins is correctly recognized. Therefore the total accuracy is $Q = C/N$.

Table 2 Comparison among different methods

Method	Recognition ratio
SVM [5]	56.0%
HKNN [16]	57.4%
RS1_HKNN_K25 [15]	60.3%
MLP [5]	48.8%
This paper	48.1% − 57.7%

In this paper there is no strategy proposed to choose a set of partitions of the classes. The total number of possible partitions is: $N!/(N - \lceil N/2 \rceil)! \lceil N/2 \rceil$ and not every partition is good. Some of then are even harmful dividing samples of the same class to both subsets. See fig. 2 (class 3). In these experiments the partitions were randomly chosen, but for example the partitions with the best recognition rate on the training dataset can be selected.

The results achieved using the proposed strategies are promising. The recognition rates obtained using these algorithms (48,1% - 57,7%) are comparable to those described in literature (48.8% - 60.3%). The described intersection method provides superior multi-class classification performance. The results show that this method offer comparable recognition rates with improved speed of training and especially in the testing phase but however some extra experiments are needed to match up to other methods.

References

1. Baldi, P., Brunak, S., Chauvin, Y., Andersen, C., Nielsen, H.: Assessing the accuracy of prediction algorithms for classification: an overview. Bioinformatics 16, 412–424 (2000)
2. Chang, C.C., Lin, C.J.: LIBSVM: a library for support vector machines (2001), Software available at, http://www.csie.ntu.edu.tw/~cjlin/libsvm

3. Chung, I.F., Huang, C.D., Shen, Y.H., Lin, C.T.: Recognition of structure classification of protein folding by NN and SVM hierarchical learning architecture. In: Kaynak, O., Alpaydın, E., Oja, E., Xu, L. (eds.) ICANN 2003 and ICONIP 2003. LNCS, vol. 2714, pp. 1159–1167. Springer, Heidelberg (2003)
4. Dietterich, T.G., Bakiri, G.: Solving multiclass problems via error-correcting output codes. Journal of Artificial Intelligence Research 2, 263–286 (1995)
5. Ding, C.H., Dubchak, I.: Multi-class protein fold recognition using support vector machines and neural networks. Bioinformatics 17, 349–358 (2001)
6. Dubchak, I., Muchnik, I., Mayor, C., Dralyuk, I., Kim, S.H.: Recognition of protein fold in the context of the Structural Classification of Proteins (SCOP) classification. Proteins 35, 401–407 (1999)
7. Fei, B., Liu, J.: Binary Tree of SVM: A New Fast Multiclass Training and Classification Algorithm. IEEE Transaction on neural networks 17(3) (May 2006)
8. Hastie, T., Tibshirani, R.: Classification by pairwise coupling. Annals of Statistics 26(2), 451–471 (1998)
9. Hobohm, U., Sander, C.: Enlarged representative set of Proteins. Protein Sci. 3, 522–524 (1994)
10. Jinbai, L., Ben, F., Lihong, X.: Binary tree of Support Vector Machine in multi-class classification problem 3rd ICECE (2004)
11. Kijsirikul, B., Ussivakul, N.: Multiclass support vector machines using adaptive directed acyclic graph. In: Proceedings of IJCNN, pp. 980–985 (2002)
12. Lee, Y., Lee, C.K.: Classification of Multiple Cancer Types by Multicategory Support Vector Machines Using Gene Expression Data. Bioinformatics 19, 1132–1139 (2003)
13. Lo Conte, L., Ailey, B., Hubbard, T.J.P., Brenner, S.E., Murzin, A.G., Chotchia, C.: SCOP: a structural classification of protein database. Nucleic Acids Res. 28, 257–259 (2000)
14. Madzarov, G., Gjorgjevskij, D., Chorbev, I.: A multi-class SVM classifier utilizing decision tree. Informatica 33, 233–241 (2009)
15. Nanni, L.: A novel ensemble of classifiers for protein fold recognition. Neurocomputing 69, 2434–2437 (2006)
16. Okun, O.: Protein fold recognition with k-local hyperplane distance nearest neighbor algorithm. In: Proceedings of the Second European Workshop on Data Mining and Text Mining in Bioinformatics, Pisa, Italy, pp. 51–57 (September 24, 2004)
17. Shen, H.B., Chou, K.C.: Ensemble classifier for protein fold pattern recognition. Bioinformatics 22, 1717–1722 (2006)
18. Vapnik, V.: The Nature of Statistical Learning Theory. Springer, New York (1995)
19. Vural, V., Dy, J.G.: A hierarchical method for multi-class support vector machines. In: Proceedings of the twenty-first ICML, Banff, Alberta, Canada, July 04–08, p. 105 (2004)
20. Wang, L., Shen, X.: Multi-category support vector machines, feature selection and solution path. Statistica Sinica 16, 617–633 (2006)

Feature Selection Using Multi-Objective Evolutionary Algorithms: Application to Cardiac SPECT Diagnosis

António Gaspar-Cunha

Abstract. An optimization methodology based on the use of Multi-Objective Evolutionary Algorithms (MOEA) in order to deal with problems of feature selection in data mining was proposed. For that purpose a Support Vector Machines (SVM) classifier was adopted. The aim being to select the best features and optimize the classifier parameters simultaneously while minimizing the number of features necessary and maximize the accuracy of the classifier and/or minimize the errors obtained. The validity of the methodology proposed was tested in a problem of cardiac Single Proton Emission Computed Tomography (SPECT). The results obtained allow one to conclude that MOEA is an efficient feature selection approach and the best results were obtained when the accuracy, the errors and the classifiers parameters are optimized simultaneously.

1 Introduction

Feature selection is of crucial importance when dealing with problems with high amount of data. This importance can be due to various reasons: i) the processing of all features available can be computational infeasible; ii) the existence of high number of variables for small number of available data points can invalidate the resolution of the problem; iii) an high number of features can be redundant or irrelevant for the classification problem under study. Therefore, taking into account the large number of variables usually present, and the frequent correlation between these variables, the existence of a feature selection method able to reduce the number of features considered for analysis is of essential importance [1].

Multi Objective Evolutionary Algorithms (MOEA) is a valid and efficient method to deal with this problem. Recently, some works using this approach have been proposed. A framework for SVM based on multi-objective optimization with the aim of minimizes the risk of the classifier and the model capacity (or accuracy) was proposed by Bi [2]. An identical approach was followed by Igel [3], but now

António Gaspar-Cunha
IPC/I3N – Institute of Polymers and Composites, University of Minho,
Campus de Azurém, Guimarães, Portugal
e-mail: agc@dep.uminho.pt

M.P. Rocha et al. (Eds.): IWPACBB 2010, AISC 74, pp. 85–92, 2010.
springerlink.com   © Springer-Verlag Berlin Heidelberg 2010

the objective concerning the minimization of the risk was replaced by the minimization of the complexity of the model (i.e., the number of features). Oliveira et al. in [4] used an hierarchical MOEA operating at two levels: performing a feature selection to generate a set of classifiers (based on artificial neural networks) and selecting the best set of classifiers. Hamdani et al. in [5] optimized simultaneously the number of features and the global error obtained by a neural network classifier using the NSGA-II algorithm [6]. Both errors of type I (false positive) and type II (false negative) were taking into account individually through the application of a MOEA by Alfaro-Cid et al. [7]. MOEA were also applied in unsupervised learning. Handl and Knowles studied the problem of feature selection by formulating them as a multi-objective optimization problem [8].

The main ideas of a previous work proposed by the author were take into account [9]. It consisted in using a MOEA to accomplish simultaneously two objectives: the minimization of the number of features used and the maximization of the accuracy of the classifier used [9]. This is an important issue since parameter tuning is not an easy task [10]. In this work these ideas were extended to deal with the issue of selecting the best accuracy measures [11-13]. Thus, different accuracy measures, such as maximization of the $F_{measure}$ and the minimization of errors (type I and type II) will be tested. Also, an analysis based on ROC curves will be carried out [13]. Simultaneously, the parameters required by the classifier will be optimized. The motivation for doing this work is the development of a methodology able to deal with bigger problems like gene expression data. However, before applying the methodology to difficult problems the methodology must be tested in small and controllable problems.

This text is organized as follows. The MOEA used will be presented and described in detail in section 2. In section 3 the classification methods employed and the main accuracy measures employed will be presented and described. The methodology proposed will be applied to a case study and the results will be presented and discussed in section 4. Finally, the conclusion will be drawn in section 5.

2   Multi-Objective Evolutionary Algorithms

Due to the complexity in dealing with multiple conflicting objectives problems, MOEAs have been recognized in the last two decades as good methods to explore and find an approximation to the Pareto-optimal front. This is due to the difficulty of traditional exact methods to solve this type of problems and by their capacity to explore and combine various solutions to find the Pareto front in a single run. The Pareto front is constituted by the non-dominated solutions, i.e., the solutions that are not better neither worst than the others. Thus, a MOEA must be able to accomplish simultaneously two objectives, a homogeneous distribution of the population along the Pareto frontier in the objective domain and an improvement of the solutions along successive generations [6, 14]. The Reduced Pareto Set Genetic Algorithm with elitism (RPSGAe) is adopted here [14, 15]. This algorithm is based on the use of a clustering technique to reduce the number of solutions on the efficient frontier, which enabled simultaneously the distribution of the solutions along the entire Pareto front and the choice of the best solutions for reproduction.

Thus, both the exploration and exploitation of the search space are simultaneously taking into account. Detailed information about this algorithm can be found elsewhere [14, 15].

3 Classification Methods

The methodology proposed here consists in using a MOEA to determine the best compromise between the two and/or the three conflicting objectives. For that purpose Support Vector Machines (SVM) will be used to evaluate (or classify) the trial solutions proposed by the MOEA during the successive generations. Support Vector Machines (SVMs) are a set of supervised learning methods based on the use of a kernel, which can be applied to classification and regression. In the SVM a hyper-plane or set of hyper-planes is (are) constructed in a high-dimensional space. In this case, a good separation is achieved by the hyper-plane that has the largest distance to the nearest training data points of any class. Thus, the generalization error of the classifier is lower when this margin is larger. SVMs can be seen an extension to nonlinear models of the generalized portrait algorithm developed by Vapnik in [16]. In this work the SVM from LIBSVM was used [17].

The SVM performance depends strongly on the selection of the right kernel, as well the definition of the best kernel parameters [3]. In the present study only the C-SVC method using as kernel the Radial Basis Function (RBF) was tested [17]. Thus, two different SVM parameters are to be selected carefully: the regularization parameter (C) and the kernel parameter (γ). Another important parameter is the training method. Two different approaches were used for training the SVM, holdout and 10-fold validation. Thus two additional parameters were studied: the Learning Rate (LR) and the Training Fraction (TF). The choice of a performance metric to evaluate the learning methods is nowadays an important issue that must be carefully defined [11-13]. Some recent studies demonstrate that the use of a single measure can introduce an error on the classifier evaluation, since two type of objectives must be accomplished simultaneously, maximization of the classifier accurateness and minimization of the errors obtained [13]. The selection of the best learning algorithm to use and the best performance metric to measure the efficiency of the classifier is nowadays the subject of many studies [11, 13].

The simplest way evaluate a classifier is the use the accuracy given by the ratio between the number instances correctly evaluated and the total number of instances, i.e., $Accuracy = (TP + TN) / (TP + TN + FP + FN)$, where, TP are the positives correctly classified, TN are the negatives correctly classified, FP are the positives incorrectly classified and FN are the negative incorrectly classified. It is also important to know the level of the errors accomplished by the classifier. Two different error types can be defined, type I and type II, given respectively by: $e_I = FP/(FP + TN)$ and $e_{II} = FN/(FN + TP)$. Another traditional way to evaluate the information is using the sensitivity or recall (R) and the precision (P) of the classifier: $R = TP/(TP + FN)$ and $P = TP/(TP + FN)$. $F_{measure}$, representing the harmonic mean of R and P, is a global measure often used to evaluate the classifier: $F_{measure}$ = $(2.P.R)/(P + R)$. In order to take into account the problem of simultaneously maximize the classifier accurateness and minimize the errors obtained, ROC curves can be adopted instead [12, 13]. On a ROC graph the False Positive rate (FP_{rate}) is plotted in the X axis and the True Positive rate (TP_{rate}) is plotted on the Y axis. Thus, defining a bi-dimensional Pareto frontier where the aim is to approach the left top corns of this graph [12, 13]. The FP_{rate} is given by the error of type I (e_I) and the TP_{rate} is given by the recall (R).

4 Results and Discussion

The MOEA methodology proposed will be used in a diagnostic problem of cardiac Single Proton Emission Computed Tomography (SPECT) images [18]. Each of the patients is classified into two categories: normal and abnormal. The database of 267 SPECT image sets (patients) was processed to extract features that summarize the original SPECT images. As a result, 44 continuous feature patterns were created for each patient. The pattern was further processed to obtain 22 binary feature patterns. The aim was finding the minimum number of features while maximizing the accuracy and/or the $F_{measure}$ and minimizing the errors. The database was downloaded from the UCI Machine Learning Repository [19].

Table 1 shows the different experiments tested. Concerning the definition of the decision variables, two possibilities were considered. Initially, a pure feature selection problem was analyzed. In this case the parameters of the classifier, such as type of training and learning rate, the SVM parameters (C and γ) and the training fraction of holdout validation, were initially set. In a second approach, these parameters were also included as variables to be optimized. The range of variation allowed for these variables is shown on Table 1. The RPSGAe was applied using the following parameters: 100 generations, crossover rate of 0.8, mutation rate of 0.05, internal and external populations with 100 individuals, limits of the clustering algorithm set at 0.2 and the number of ranks (N_{Ranks}) at 30. These values resulted from a carefully analysis made previously [14, 15]. Due to the stochastic nature of the initial tentative solutions several runs have to be performed (in the present case 10 runs) for each experiment. Thus, a statistical method based on attainment functions was applied to compare the final population for all runs [20, 21]. This method attributes to each objective vector a probability that this point is attaining in one single run [20]. It is not possible to compute the true attainment function, but it can be estimated based upon approximation set samples, i.e., different approximations obtained in different runs, which is denoted as Empirical Attainment Function (EAF) [21]. The differences between two algorithms can be visualized by plotting the points in the objective space where the differences between the empirical attainment functions of the two algorithms are significant [22].

Table 1 Experiments

Exp.	γ	C	TM	LR	TF	Objectives
H01	10	1	K(10)	0.01	*	NA + PA
H02	10	1	K(10)	0.01	*	$NA + e_I$
H03	10	1	K(10)	0.01	*	$NA + e_{II}$
H04	10	1	K(10)	0.01	*	$NA + F_m$
H05	10	1	K(10)	0.01	*	$NA + e_I + F_m$
H06	[0.01,10]	[1,150]	K(10)	[0.001,0.1]	*	$NA + F_m$
H07	[0.01,10]	[1,150]	K(10)	[0.001,0.1]	*	$NA + e_I + F_m$
H08	10	1	H	0.01	0.7	$NA + F_m$
H09	[0.01,10]	[1,150]	H	[0.001,0.1]	[0.2,0.9]	$NA + F_m$
H10	[0.01,10]	[1,150]	H	[0.001,0.1]	[0.2,0.9]	$NA + e_I + F_m$
H11	10	1	K(10)	0.01	*	$NA + e_I + R$
H12	[0.01,10]	[1,150]	K(10)	[0.001,0.1]	*	$NA + e_I + R$
H13	[0.01,10]	[1,150]	K(10)	[0.001,0.1]	*	$NA + e_I + R + F_m$

* Not applicable.

Figure 1 shows the initial population and the Pareto front after 100 generations for the first run of Experiments H01 and H02 (Table 1). Identical results were obtained for the remaining runs. As can be observed there is a clear improvement of the solutions proposed during the search process. The algorithm was able to evolve to good values of the *Accuracy* (graph at the left) using a few features. In fact only six or seven features are needed to reach more than 90% of accuracy. Concerning the experiments were the e_I was minimized simultaneously with the number of features (H02) identical improvements can be noticed. More results can be found at http://www.dep.uminho.pt/agc/agc/Supplementary_Information_Page.html.

The results for the first run of experiment 5 were plotted in Figure 2. In this case a 3-dimensional Pareto front was obtained and some of the points that seem to be dominated in one of the graphs (in each 2D plots) are in reality non-dominated due to the third objective considered in the optimization run. These plots are very similar to those obtained for experiments H01 and H02, but now the solutions resulted from a compromise between the 3 objectives considered simultaneously. Thus, more features are needed to satisfy simultaneously the maximization of F_{measue} and the minimization of e_I. These plots allow us to observe the shape of the curves and to get some information about the relation between the objectives. This information is important in the sense that will help the decision maker selecting the best solution satisfying their requirements.

The EAFs functions were used to compare experiments H04, H06, H08 and H09 (due to a lack of space these results were presented in the supplementary information page identified above). This analysis allowed concluding that the best performance is obtained with the k-fold validation method when the classifier parameters are optimized simultaneously (experiment H06). Finally, the advantages of using the proposed methodology for dealing with this type of problems, is shown in Figure 3 (Pareto fronts for experiment H13). In this figure is possible to

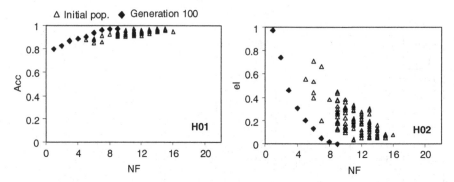

Fig. 1 Pareto fronts after 100 generations for runs H01 and H02 of table 1

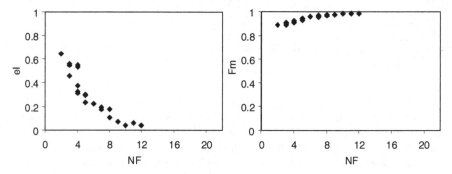

Fig. 2 Three-dimensional Pareto fronts after 100 generations for run H05

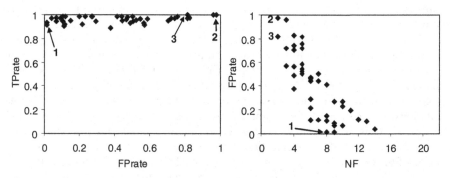

Fig. 3 Pareto fronts for experiment H13 (at left is possible to observe the solutions generated in the ROC curve domain)

observe that the algorithm is able to converge to very good solutions since high values for TP_{rate} were obtained simultaneously with low values for FP_{rate}. This indicates that the application of a MOEA, where the features to be selected and the parameters of the SVM are optimized simultaneously, is a method with good potentialities for solving this type of problems. The solutions identified in this plots are presented in Table 2. These include the decision variables (features selected, and classifier parameters) and the objective values. Identical results were obtained for runs H11 and H12.

Table 2 Best solutions for the first run of experiment H13 and identified in Figure 3

Sol.	Decision Variables				Objectives			
	Features selected	γ	C	LR	NF	$FPrate$	$TPrate$	F_m
1	F3,F4,F11,F13,F14,F16,F18,F22	0.078	0.17	62.7	8	0.013	0.91	0.951
2	F4, F11	0.040	0.43	78.7	2	0.975	1.00	0.886
3	F11, F13	0.043	0.46	81.7	2	0.818	0.97	0.892

5 Conclusions

In this work a MOEA was used for feature selection in data mining problems using a Support Vector Machines classifier. The methodology proposed was able not only to propose solutions with a few number of features necessary but is able also to provide relevant information to the decision maker, such as the best features to be used but, the best parameters of the classifier and the trade-off between the different objectives used. Finally, the approach followed here showed good potentialities in obtaining a good approximation to the ROC curves.

References

1. Guyon, I., Gunn, S., Nikravesh, M., Zadeh, L.: Feature Extraction Foundations and Applications. Springer, Heidelberg (2006)
2. Bi, J.: Multi-Objective Programming in SVMs. In: Proceedings of the Twentieth International Conference on Machine Learning, ICML-2003, Washington DC (2003)
3. Igel, C.: Multi-Objective Model Selection for Support Vector Machines. In: Coello Coello, C.A., Hernández Aguirre, A., Zitzler, E. (eds.) EMO 2005. LNCS, vol. 3410, pp. 534–546. Springer, Heidelberg (2005)
4. Oliveira, L.S., Morita, M., Sabourin, R.: Feature Selection for Ensembles Using the Multi-Objective Optimization Approach. Studies in Computational Intelligence 16, 49–74 (2006)
5. Hamdani, T.M., Won, J.M., Alimi, A.M., Karray, F.: Multi-objective Feature Selection with NSGA II. In: Beliczynski, B., Dzielinski, A., Iwanowski, M., Ribeiro, B. (eds.) ICANNGA 2007. LNCS, vol. 4431, pp. 240–247. Springer, Heidelberg (2007)
6. Deb, K., Pratap, A., Agarwal, S., Meyarivan, T.: A fast and elitist multi-objective genetic algorithm: NSGA-II. IEEE Transaction on Evolutionary Computation 6, 181–197 (2002)
7. Alfaro-Cid, E., Castillo, P.A., Esparcia, A., Sharman, K., Merelo, J.J., Prieto, A., Mora, A.M., Laredo, J.L.: Comparing Multiobjective Evolutionary Ensembles for Minimizing Type I and II Errors for Bankruptcy Prediction. In: Congress on Evolutionary Computation - CEC 2008, Washington, USA, pp. 2907–2913 (2008)

8. Handl, J., Knowles, J.: Feature subset selection in unsupervised learning via multiobjective optimization. International Journal of Computational Intelligence Research 2, 217–238 (2006)
9. Mendes, F., Duarte, J., Vieira, A., Ribeiro, B., Ribeiro, A., Neves, J., Gaspar-Cunha, A.: Feature Selection for Bankruptcy Prediction: A Multi-Objective Optimization Approach. International Journal of Natural Computing Research (2010) (submitted for publication)
10. Kulkarni, A., Jayaraman, V.K., Kulkarni, V.D.: Support vector classification with parameter tuning assisted by agent-based technique. Comp. and Chemical Engineering 28, 311–318 (2004)
11. Caruana, R., Niculescu-Mizil, A.: Data Mining in Metric Space: An Empirical Analysis of Supervised Learning Performance Criteria, in Proc. Internat. In: Proc. Internat. Conf. on Knowledge Discovery and Data Mining, Seattle, Washington, pp. 69–78 (2004)
12. Provost, F., Fawcet, T.: Analysis and Verification of Classifier Performance: Classification under Imprecise Class and Cost Distributions. In: Proc. Internat. Conf. on Knowledge Discovery and Data Mining, KDD 1997, Menlo Park, CA, pp. 43–48 (1997)
13. Fawcet, T.: An introduction to ROC analysis. Pattern Recognition Letters 27, 861–874 (2006)
14. Gaspar-Cunha, A., Covas, J.A.: - RPSGAe - A Multiobjective Genetic Algorithm with Elitism: Application to Polymer Extrusion. In: Jorrand, P., Kelemen, J. (eds.) FAIR 1991. LNCS, vol. 535, pp. 221–249. Springer, Heidelberg (1991)
15. Gaspar-Cunha, A.: Modelling and Optimization of Single Screw Extrusion, Published doctoral dissertation. In: Gaspar-Cunha, A., Köln (eds.) Modelling and Optimization of Single Screw Extrusion: Using Multi-Objective Evolutionary Algorithms. Lambert Academic Publishing, Germany (2000)
16. Cortes, C., Vapnik, V.: Support-Vector Networks. Machine Learning 20, 273–297 (1995)
17. Chang, C.C., Lin, C.J.: LIBSVM a library for support vector machines (Tech. Rep.), Department of Computer Science and Information Engineering, National Taiwan University, Taipei, Taiwan (2000)
18. Kurgan, L.A., Cios, K.J., Tadeusiewicz, R., Ogiela, M., Goodenday, L.S.: Knowledge Discovery Approach to Automated Cardiac SPECT Diagnosis. Artificial Intelligence in Medicine 23, 149–169 (2001)
19. Asuncion, A., Newman, D.J.: UCI Machine Learning Repository. University of California, School of Information and Computer Science, Irvine (2007), http://www.ics.uci.edu/~mlearn/MLRepository.html (Accessed 10 February 2010)
20. Fonseca, C., Fleming, P.J.: On the performance assessment and comparison of stochastic multiobjective optimizers. In: Ebeling, W., Rechenberg, I., Voigt, H.-M., Schwefel, H.-P. (eds.) PPSN 1996. LNCS, vol. 1141, pp. 584–593. Springer, Heidelberg (1996)
21. Fonseca, V.G., Fonseca, C., Hall, A.: Inferential performance assessment of stochastic optimisers and the attainment function. In: Zitzler, E., Deb, K., Thiele, L., Coello Coello, C.A., Corne, D.W. (eds.) EMO 2001. LNCS, vol. 1993, pp. 213–225. Springer, Heidelberg (2001)
22. López-Ibañez, M., Paquete, L., Stützle, T.: Hybrid population based algorithms for the bi-objective quadratic assignment problem. Journal of Mathematical Modelling and Algorithms 5, 111–137 (2006)

Two Results on Distances for Phylogenetic Networks

Gabriel Cardona, Mercè Llabrés, and Francesc Rosselló

Abstract. We establish two new results on distances for phylogenetic networks. First, that Nakhleh's metric for reduced networks [15] is a metric on the class of semibinary tree-sibling networks, being thus the only non-trivial metric computable in polynomial-time available so far on this class. And second, that a simple generalization of the nodal distance from binary phylogenetic trees to binary phylogenetic networks has the same separation power as the one given in [8].

1 Introduction

Phylogenetic networks have been studied over the last years as a richer model of the evolutionary history of sets of organisms than phylogenetic trees, because they take into account not only mutation events but also reticulation events, like recombinations, hybridizations, and lateral gene transfers. Technically, it is accomplished by modifying the concept of phylogenetic tree in order to allow the existence of nodes with in-degree greater than one. As a consequence, much progress has been made to find practical algorithms for reconstructing a phylogenetic network from a set of sequences or other types of evolutive information. Since different reconstruction methods applied to the same sequences, or a single method applied to different sequences, may yield different phylogenetic networks for a given set of species, a sound measure to compare phylogenetic networks becomes necessary [13]. The comparison of phylogenetic networks is also needed in the assessment of phylogenetic reconstruction methods [12], and it will be required to perform queries on future databases of phylogenetic networks [17].

Several distances for the comparison of phylogenetic networks have been proposed so far in the literature, including generalizations to networks of the Robinson-Foulds distance for trees, like the tripartitions distance [13] and the μ-distance [2, 9],

Gabriel Cardona · Mercè Llabrés · Francesc Rosselló
Department of Math & Computer Science, University of Balearic Islands, Crta. Valldemossa, km. 7,5, 07122 Palma de Mca, Spain
e-mail: {gabriel.cardona,merce.llabres,cesc.rossello}@uib.es

M.P. Rocha et al. (Eds.): IWPACBB 2010, AISC 74, pp. 93–100, 2010.
springerlink.com   © Springer-Verlag Berlin Heidelberg 2010

and different types of nodal distances [4, 8]. All polynomial time computable distances for phylogenetic networks introduced up to now do not separate arbitrary phylogenetic networks, that is, zero distance does not imply in general isomorphism. Of course, this is consistent with the equivalence between the isomorphism problems for phylogenetic networks and for graphs, and the general belief that the latter lies in NP−P. Therefore one has to study for which interesting classes of phylogenetic networks these distances are metrics in the precise mathematical sense of the term. The interest of the classes under study may stem from their biological significance, or from the existence of reconstruction algorithms.

This work contributes to this line of research in two aspects. On the one hand, we prove that a distance introduced recently by Nakhleh [15] separates semibinary tree-sibling phylogenetic networks (roughly speaking, networks where every node of in-degree greater than one has in-degree exactly two and at least one sibling of in-degree 1; see the next section for the exact definition, [2] for a discussion of the biological meaning of this condition, and [11, 16] for reconstruction algorithms). In this way, this distance turns out to be the only non-trivial metric available so far on this class of networks that is computable in polynomial-time. On the other hand, we propose a simple generalization of the nodal, or path-difference, distance for binary phylogenetic trees to binary phylogenetic networks that turns out to have the same separation power, up to the current state of knowledge, as the more involved generalization introduced in [8], but smaller computation time.

2 Preliminaries

Given a set S of *labels*, a *S-DAG* is a directed acyclic graph with its leaves bijectively labelled by S. In a S-DAG, we shall always identify without any further reference every leaf with its label.

Let $N = (V, E)$ be a S-DAG, with V its set of nodes and E its set of edges. A node is a *leaf* if it has out-degree 0 and *internal* otherwise, a *root* if it has in-degree 0, of *tree* type if its in-degree is ≤ 1, and of *hybrid* type if its in-degree is > 1. N is *rooted* when it has a single root. A node v is a *child* of another node u (and hence u is a *parent* of v) if $(u, v) \in E$. Two nodes with a parent in common are *sibling* of each other. A node v is a *descendant* of a node u when there exists a path from u to v: we shall also say in this case that u is an *ancestor* of v. The *height* $h(v)$ of a node v is the largest length of a path from v to a leaf.

A *phylogenetic network* on a set S of *taxa* is a rooted S-DAG such that no tree node has out-degree 1 and every hybrid node has out-degree 1. A *phylogenetic tree* is a phylogenetic network without hybrid nodes.

The underlying biological motivation for these definitions is that tree nodes model species (either extant, the leaves, or non-extant, the internal tree nodes), while hybrid nodes model reticulation events. The parents of a hybrid node represent the species involved in this event and its single child represents the resulting species (if it is a tree node) or a new reticulation event where this resulting species gets involved into without yielding any other descendant (if the child is a hybrid node).

The tree children of a tree node represent direct descendants through mutation. The absence of out-degree 1 tree nodes in phylogenetic network means that every non-extant species has at least two different direct descendants. This is a very common restriction in any definition of phylogeny, since species with only one child cannot be reconstructed from biological data.

Many restrictions have been added to this definition. Let us introduce now some of them. For more information on these restrictions, including their biological or technical motivation, see the references accompanying them.

- A phylogenetic network is *semibinary* if every hybrid node has in-degree 2 [2], and *binary* if it is semibinary and every internal tree node has out-degree 2.
- A phylogenetic network is *tree-child* (*TC*) [9] if every internal node has a child that is a tree node, and it is *tree-sibling* (*TS*) [2, 14] if every hybrid node has a sibling that is a tree node.
- A phylogenetic network is a *galled tree* [10] when no node belongs to two different reticulation cycles (a *reticulation cycle* is a pair of paths with the same origin and end, and disjoint sets of intermediate nodes). Every galled tree is TC and semibinary [18].
- A phylogenetic network is *time consistent* [1] when it admits a *time assignment*, that is, a mapping $\tau : V \to \mathbb{N}$ such that:
    - if $(u,v) \in E$ and v is a hybrid node, then $\tau(u) = \tau(v)$,
    - if $(u,v) \in E$ and v is a tree node, then $\tau(u) < \tau(v)$.

    Time consistent tree-child and tree-sibling phylogenetic networks will be denoted, for simplicity, TCTC and TSTC, respectively.
- A phylogenetic network is *reduced* [13] when the only pairs of nodes with the same sets of descendant leaves consist of a hybrid node and its only child, provided that the latter is of tree type.

3   On Nakhleh's Distance m

Let us recall the distance m introduced by Nakhleh in [14], in the version described in [5]. Let $N = (V,E)$ be a phylogenetic network on a set S of taxa. For every node $v \in V$, its *nested label* $\lambda_N(v)$ is defined by recurrence as follows:

- If v is the leaf labelled i, then $\lambda_N(v) = \{i\}$.
- If v is internal and all its children $v_1, \ldots, v_k$ have been already labelled, then $\lambda_N(v)$ is the multiset $\{\lambda_N(v_1), \ldots, \lambda_N(v_k)\}$ of their labels.

The absence of cycles in N entails that this labelling is well-defined.

The *nested labels representation* of N is the multiset

$$\lambda(N) = \{\lambda_N(v) \mid v \in V\},$$

where each nested label appears with multiplicity the number of nodes having it as nested label. *Nakhleh's distance m* between a pair of phylogenetic networks N, N' on a same set S of taxa is then

$$m(N,N') = |\lambda(N) \triangle \lambda(N')|,$$

where the symmetric difference and the cardinal refer to multisets.

This distance trivially satisfies all axioms of metrics except, at most, the separation axiom, and thus this is the key property that has to be checked on some class of networks in order to guarantee that m is a metric on it. So far, this distance m is known to be a metric for reduced networks [15], TC networks [5], and semibinary TSTC networks [5] (always on any fixed set of labels S). On the other hand, it is not a metric for arbitrary TSTC networks [5]. We add now the following result.

Theorem 1. *Nakhleh's distance m is a metric on the class of all semibinary TS networks on any fixed set of labels S.*

Proof. Fisrt of all, we prove that two different nodes in a semibinary TS (*sbTS*, for short) network have always different nested labels. Indeed, assume that v, w are two nodes in a sbTS network N such that $\lambda_N(w) = \lambda_N(v)$. Without any loss of generality, we assume that v is a node of smallest height with the same nested label as another node. By definition, v cannot be a leaf, because the only node with nested label $\{i\}$, with $i \in S$, is the leaf labelled i. Therefore v is internal: let $v_1,\ldots,v_k$ ($k \geqslant 1$) be its children, so that $\lambda_N(v) = \{\lambda_N(v_1),\ldots,\lambda_N(v_k)\}$. Since $\lambda_N(w) = \lambda_N(v)$, w has k children, say $w_1,\ldots,w_k$, and they are such that $\lambda_N(v_i) = \lambda_N(w_i)$ for every $i = 1,\ldots,k$. Then, since $v_1,\ldots,v_k$ have smaller height than v and by assumption v is a node of smallest height among those nodes with the same nested label as some other node, we deduce that $v_i = w_i$ for every $i = 1,\ldots,k$. Therefore, $v_1,\ldots,v_k$ are hybrid, and their parents v, w (which are their only parents, by the semibinarity condition) have no more children, which implies that $v_1,\ldots,v_k$ do not have any tree sibling. This contradicts the TS condition for N.

The non existence of any pair of different nodes with the same nested label in a sbTS network $N = (V,E)$ implies that, for every $u, v \in V$, $(u,v) \in E$ iff $\lambda_N(v) \in \lambda_N(u)$ (indeed: on the one hand, the very definition of nested label entails that if $(u,v) \in E$, then $\lambda_N(v) \in \lambda_N(u)$; and conversely, if $\lambda_N(v) \in \lambda_N(u)$, then u has a child v' such that $\lambda_N(v') = \lambda_N(v)$, and by the injectivity of nested labels, it must happen that $v = v'$). This clearly implies that a sbTS network can be reconstructed, up to isomorphisms, from its nested labels representation, and hence that non-isomorphic sbTS networks always have different nested label representations.   □

4   A Simpler Nodal Distance for Binary Networks

For simplicity, assume in this section that S is a finite set of non-negative integers. The *path-length* $L_T(i,j)$ between a pair of leaves $i, j \in S$ in a phylogenetic tree T on S is the length of the unique undirected path connecting them. It is well known that the *path-lengths vector* $(L_T(i,j))_{i,j \in S,\ i<j}$ characterizes up to isomorphism a binary phylogenetic tree, but there exist non-isomorphic non-binary phylogenetic trees with the same path-lengths vectors [7]. This means that any distance for phylogenetic trees based on the comparison of these vectors will only be a metric for *binary* trees.

In [8] we generalized the notion of path-length between a pair of leaves from phylogenetic trees to phylogenetic networks, by defining the *LCSA-path length* between two leaves as the sum of the distances from the leaves to their *Least Common Semistrict Ancestor* [3, §IV], a certain uniquely defined common ancestor of them that plays the role in phylogenetic networks of the least common ancestor in phylogenetic trees (two leaves in a phylogenetic network need not have a least common ancestor). In this section we use a simpler approach to define a path-length between a pair of leaves. Actually, the simplest way would be to define the (*undirected*) *path-length* between a pair of leaves as the length of a shortest undirected path connecting them. But these path-lengths do not characterize a binary phylogenetic network, even in the most restrictive case: time consistent galled trees (see Fig. 1). Therefore, we must take here another approach.

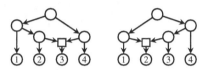

Fig. 1 Two non-isomorphic binary time consistent galled trees with the same undirected path-lengths between pairs of leaves

For every pair of leaves $i, j \in S$ of a phylogenetic network N on S, let the *TCA-path-length* (TCA stands for 'Through a Common Ancestor') $\ell_N(i, j)$ between them be the length of a shortest undirected path consisting of a pair of paths from a common ancestor of i and j to them. Of course, if N is a phylogenetic tree, then the TCA-path-length is equal to the usual path-length.

The *TCA-path-lengths vector* of N is $\ell(N) = (\ell_N(i, j))_{i,j \in S, \, i<j}$. Since the path-lengths vectors only separate binary phylogenetic trees, it only makes sense to ask in which classes of *binary* phylogenetic networks do the TCA-path-lengths vectors separate its members. We have the following result:

Theorem 2. *The TCA-path-lengths vectors separate binary TCTC networks, but they do not separate binary galled trees, TSTC networks, or TC networks.*

The counterexamples corresponding to the negative part of this statement can be found in Fig. 2. We now sketch a proof of the positive assertion on binary TCTC networks (*bTCTC* networks, for short), which is based on algebraic induction using reductions (for a detailed explanation of this technique, see [6, §V]). This proof is similar to the proof given in [8] of the same property for LCSA-path-length vectors.

To begin with, we introduce a pair of reduction procedures that decrease the number of leaves and nodes in a bTCTC network.

(R) Let i, j be two sibling leaves in a bTCTC network N, and let v be their parent. The R($i; j$) *reduction* of N is the bTCTC network $N_{R(i;j)}$ obtained by removing the leaves i, j from N, together with their incoming arcs, and labelling with j the node v (which has become a leaf); cf. Fig. 3.

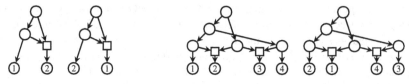

Fig. 2 Left: Two non-isomorphic binary galled trees (and in particular, TC networks) with the same vectors of TCA-distances. Right: Two non-isomorphic binary TSTC networks with the same vectors of TCA-distances

(H) Let A be a hybrid node in a bTCTC network N such that its only child is the leaf i and its siblings are the leaves j_1, j_2; let v_1 and v_2 be, respectively, the parents of these leaves. The H$(i; j_1, j_2)$ *reduction* of N is the bTCTC network $N_{H(i;j_1,j_2)}$ obtained by removing the nodes i, j_1, j_2, A, together with their incoming arcs, and labelling the nodes v_1 and v_2 (which have become leaves) with j_1 and j_2, respectively; cf. Fig. 3.

Fig. 3 Left: The R$(i; j)$ reduction. Right: The H$(i; j_1, j_2)$ reduction

Let $R^{-1}(i; j)$ and $H^{-1}(i; j_1, j_2)$ be the inverse expansions of R$(i; j)$ and H$(i; j_1, j_2)$, respectively. Given a bTCTC network N without any leaf labelled with i:

- The $R^{-1}(i; j)$ expansion can be applied to N if it has a leaf labelled with j, and then $N_{R^{-1}(i;j)}$ is the bTCTC network obtained by removing this label, adding two tree children to the corresponding node, and labelling them with i and j.
- The $H^{-1}(i; j_1, j_2)$ expansion can be applied to N if it contains a pair of leaves labelled with j_1 and j_2 with the same time value under some time assignment, and then $N_{H^{-1}(i;j_1,j_2)}$ is the bTCTC network obtained by first removing the labels j_1 and j_2, and then, if we denote by v_1 and v_2 the resulting nodes, by adding a new (hybrid) node A, three tree leaves labelled with i, j_1 and j_2, and arcs $(v_1, j_1), (v_2, j_2), (v_1, A), (v_2, A), (A, i)$.

It is straightforward to check that an expansion $R^{-1}(i; j)$ or $H^{-1}(i; j_1, j_2)$ can always be applied to the result of an application of the respective R$(i; j)$ or H$(i; j_1, j_2)$ reduction, and that the application of these expansions preserve isomorphisms. Moreover, arguing as in [8, Prop. 17], we can prove the following result.

Proposition 1. *Let N be a bTCTC network with more than one leaf. Then, there always exist some R or H reduction that can be applied to N.*

So, by [6, Lem. 6], and since there is only one bTCTC network with 1 leaf (up to relabelling and isomorphisms), to prove that the TCA-path lengths vectors separate bTCTC networks, it is enough to prove that the possibility of applying a reduction to a bTCTC network N can be decided from $\ell(N)$, and that the TCA-path-lengths vector of the result of the application of a reduction to a bTCTC network N depends only on $\ell(N)$ and the reduction. These two facts are given by the the following lemma. Its proof is similar to (and easier than) those of Lemmas 20 and 22 in [8], and we do not include it.

Lemma 1. *Let N be a bTCTC network on a set S.*

(1) $R(i;j)$ can be applied to N iff $\ell_N(i,j) = 2$. And if $R(i;j)$ can be applied to N, then

$$\ell_{N_{R(i;j)}}(j,k) = \ell_N(j,k) - 1 \quad \text{for every } k \in S \setminus \{i,j\}$$
$$\ell_{N_{R(i;j)}}(k,l) = \ell_N(k,l) \quad \text{for every } k,l \in S \setminus \{i,j\}$$

(2) $H(i;j_1,j_2)$ can be applied to N iff

- $\ell_N(i,j_1) = \ell_N(i,j_2) = 3$,
- *if $\ell_N(j_1,j_2) = 4$, then $\ell_N(j_1,k) = \ell_N(j_2,k)$ for every $k \in S \setminus \{j_1,j_2,i\}$.*

And if $H(i;j_1,j_2)$ can be applied to N, then

$$\ell_{N_{H(i;j_1,j_2)}}(j_1,j_2) = \ell_N(j_1,j_2) - 2$$
$$\ell_{N_{H(i;j_1,j_2)}}(j,k) = \ell_N(j,k) - 1 \text{ for every } k \in S \setminus \{i,j_1,j_2\} \text{ and } j \in \{j_1,j_2\}$$
$$\ell_{N_{H(i;j_1,j_2)}}(k,l) = \ell_N(k,l) \text{ for every } k,l \in S \setminus \{i,j_1,j_2\}$$

We define the *TCA-nodal distance* between a pair of phylogenetic networks N,N' as the Manhattan distance between their TCA-path-lengths vectors:

$$d_{TCA}(N,N') = \sum_{i,j \in S,\ i<j} |\ell_N(i,j) - d_{N'}(i,j)|.$$

This distance satisfies trivially all axioms of metrics except, at most, the separation axiom. Therefore, Thm. 2 implies the following result.

Corollary 1. *The TCA-nodal distance is a metric on the class of all bTCTC networks on a given set S of taxa, but it is not a metric for binary galled trees, TSTC networks, or TC networks.*

Of course, we could have defined a nodal distance for phylogenetic networks by comparing their TCA-path lengths vectors using any other metric for real-valued vectors, for instance the euclidean metric, and the conclusion would be the same.

To frame this corollary, recall that the *LCSA-nodal distance* defined by comparing the LCSA-path-lengths vectors using the Manhattan distance satisfies exactly the same result: it is a metric for bTCTC networks, but not for binary galled trees, TSTC networks, or TC networks [6, 8]. On the other hand, the LCSA-nodal distance between a pair of phylogenetic networks with n leaves, m internal nodes and e arcs can be computed in time $O(me + n^2m)$ using a simple algorithm (cf. [4]). The natural adaptation of this algorithm to the computation of the TCA-nodal distance runs in time only $O(ne + n^2m)$.

Acknowledgements. This work has been partially supported by the Spanish Government and FEDER funds, through project MTM2009-07165 and TIN2008-04487-E/TIN.

References

1. Baroni, M., Semple, C., Steel, M.: Hybrids in real time. Syst. Biol. 55, 46–56 (2006)
2. Cardona, G., Llabrés, M., Rosselló, F., Valiente, G.: A distance metric for a class of tree-sibling phylogenetic networks. Bioinformatics 24, 1481–1488 (2008)
3. Cardona, G., Llabrés, M., Rosselló, F., Valiente, G.: Metrics for phylogenetic networks I: Generalizations of the Robinson-Foulds metric. IEEE T. Comput. Biol. 6, 1–16 (2009)
4. Cardona, G., Llabrés, M., Rosselló, F., Valiente, G.: Metrics for phylogenetic networks II: Nodal and triplets metrics. IEEE T. Comput. Biol. 6, 454–469 (2009)
5. Cardona, G., Llabrés, M., Rosselló, F., Valiente, G.: On Nakhleh's metric for reduced phylogenetic networks. IEEE T. Comput. Biol. 6, 629–638 (2009)
6. Cardona, G., Llabrés, M., Rosselló, F., Valiente, G.: Comparison of galled trees. IEEE T. Comput. Biol. (in press, 2010)
7. Cardona, G., Llabrés, M., Rosselló, F., Valiente, G.: Nodal distances for rooted phylogenetic trees. J. Math. Biol. (in press, 2010)
8. Cardona, G., Llabrés, M., Rosselló, F., Valiente, G.: Path lengths in tree-child time consistent hybridization networks. Information Sciences 180, 366–383 (2010)
9. Cardona, G., Rosselló, F., Valiente, G.: Comparison of tree-child phylogenetic networks. IEEE T. Comput. Biol. 6, 552–569 (2009)
10. Gusfield, D., Eddhu, S., Langley, C.H.: Optimal, efficient reconstruction of phylogenetic networks with constrained recombination. J. Bioinform. Comput. Biol. 2(1), 173–213 (2004)
11. Jin, G., Nakhleh, L., Snir, S., Tuller, T.: Maximum likelihood of phylogenetic networks. Bioinformatics 22(21), 2604–2611 (2006)
12. Moret, B.M.E.: Computational challenges from the tree of life. In: Demetrescu, C., Sedgewick, R., Tamassia, R. (eds.) Proc. 7th Workshop Algorithm Engineering and Experiments and 2nd Workshop Analytic Algorithmics and Combinatorics, pp. 3–16. SIAM, Philadelphia (2005)
13. Moret, B.M.E., Nakhleh, L., Warnow, T., Linder, C.R., Tholse, A., Padolina, A., Sun, J., Timme, R.: Phylogenetic networks: Modeling, reconstructibility, and accuracy. IEEE T. Comput. Biol. 1, 13–23 (2004)
14. Nakhleh, L.: Phylogenetic networks. Ph.D. thesis, University of Texas at Austin (2004)
15. Nakhleh, L.: A metric on the space of reduced phylogenetic networks. IEEE T. Comput. Biol. (in press, 2010)
16. Nakhleh, L., Warnow, T., Linder, C.R., John, K.S.: Reconstructing reticulate evolution in species: Theory and practice. J. Comput. Biol. 12(6), 796–811 (2005)
17. Page, R.D.M.: Phyloinformatics: Toward a phylogenetic database. In: Wang, J.T.L., Zaki, M.J., Toivonen, H., Shasha, D. (eds.) Data Mining in Bioinformatics, pp. 219–241. Springer, Heidelberg (2005)
18. Rosselló, F., Valiente, G.: All that glisters is not galled. Mathematical Biosciences 221, 54–59 (2009)

Cramér Coefficient in Genome Evolution

Vera Afreixo and Adelaide Freitas

Abstract. The knowledge of full genome sequences for a growing number of organisms and its accessibility in Web has motivated an increasing development of both statistical and bioinformatics tools in Genetics. One of main goals is the identification of relevant statistical structures in DNA in order to understand the general rules that govern gene primary structure organization and its evolution.

In this paper, we analyze the dependence between genetic symbols for the total coding sequences of twelve species. For each specie we define codon bias profiles calculating the levels of association between codons spaced by d codons and the levels of association between codons and nucleotides spaced by d nucleotides, $d \in \{0, 1, 2, \ldots\}$, given by the Cramér coefficient usually applied in contingency tables. Comparisons among the twelve species suggest these profiles can capture essential information about DNA structure in a genomic scale allowing the construction of a dendrogram which is, in some aspects, coherent with the biologically evolution for these species.

1 Introduction

The DNA coding sequences are composed by sequential non-random associations of 4 nucleotides (Adenine – A, Cytosine – C, Guanine – G, and Thymine – T). The number of nucleotides in each coding sequence is multiple of 3 and each triple of nucleotides in a specific frame define one codon. Each coding sequence is initialized with the codon ATG and finished with the codon TAA, TAG or TGA (stop codons).

Vera Afreixo · Adelaide Freitas
Department of Mathematics, University of Aveiro,
Campus de Santiago 3810-193 Aveiro, Portugal
e-mail: vera@ua.pt, adelaide@ua.pt

M.P. Rocha et al. (Eds.): IWPACBB 2010, AISC 74, pp. 101–107, 2010.
springerlink.com

There are $4^3 = 64$ possible codons codifying only 20 proteinic units or amino acids and 3 stop codons that indicate that the protein in construction is concluded. This redundancy has been studied in different manners in order to unveil gene primary structure features: codon usage, codon pair context, N_1 context, codon adaptation index, among others (see, for example, [2, 3, 4, 5, 8, 10, 11, 12]).

Also distinct statistical methodologies have been applied to test the existence of bias between genetics symbols. Some of these methodologies are based on statistic test that compares the observed frequencies with the expected frequencies under independence between the variables under study. A ratio called relative abundance is suggested in [5]. The authors studied the influence of neighbor nucleotides surrounding each codon belonging to a pre-defined synonymous group. For codon pair context, in [7, 9] is proposed an analysis of adjusted residuals based on Pearson's chi-squared statistic after the construction of the contingency table obtained by counting all combinations of two consecutive codons from the sequenced genome. The statistical analysis confirmed the existence of pairs highly biased. One advantage of this methodology is the existence of complementary tools such as measures of association to quantify the strength of this dependence (bias).

Herein we propose to evaluate the DNA dependence structure through codon bias profiles given by the degrees of association between codons spaced by d codons and the degrees of association between codons and nucleotides spaced by d nucleotides, $d \in \{0, 1, 2, \ldots\}$, calculated by the Cramér coefficient usually applied as a measure of association in contingency tables. The application of this methodology on fully sequenced genomes of 2 Bacteria and 10 Eukaryota provided a tree that describes an expectable phylogenetic evolution of these species. Phylogenetic trees reproduce evolutionary trees that represent the historical relationships between the species. Recent phylogenetic tree algorithms use nucleotide sequences. Typically, these trees are constructed on multiple sequence alignments of homologous genes [6], which is a computationally demanding task. Our proposal is based on global patterns of the association of codons and codon-nucleotide pairs in the total coding sequences in each specie. Our results on the twelve species suggest that an analysis of the Cramér coefficient can shed new light into important gene primary structure features and provide an algorithm based on association values between spaced codons that is able to build a kind of phylogenetic trees.

For easy computation we use *Anaconda* (http://bioinformatics.ua.pt/applications/anaconda) which is a bioinformation system for automated statistical analysis of coding sequences on a genomic scale [9]. In Section 2 we descreve the procedures in order to obtain the codon bias profiles in terms of the levels of association between genetic symbols. In Section 3 we present the experimental results of our methodology when applied to twelve species, summarizing our conclusions and future research in Section 4.

2 Methods

Given a completely sequenced genome, the bioinformatics system *Anaconda* identifies the start codon and reads the complete gene sequence of an genome fixing a codon (ribosomal P-site) and memorizing its neighbors codons (E-site codon and A-site codon) [9]. The data are then processed into several contingency tables of variable pairs: (*amino acid, amino acid*), (*codon, codon*), (*codon, amino acid*), and depending on the position of a nucleotide in the codon, (*codon, first nucleotide*), (*codon, second nucleotide*) and (*codon, third nucleotide*), considering both 3' and 5' reading directions by the ribosome, and where the genetic symbols in each pair are spaced by d codons in gene sequences, for any choice of $d \in \{0, 1, 2, \ldots\}$ input by user [9]. For instance, for the pair (*codon, codon*) with $d = 0$ and 3' direction *Anaconda* provides a 61×64 contingency table of consecutive codon pairs, where the 61 rows correspond to the all possible codons in the P-site and the 64 columns to the all codons in the A-site of the ribosome. For the contexts involving one nucleotide, 61×4 contingency tables will be obtained.

For testing the independence between the two variables in each table, the Pearson chi-squared statistic (χ^2) is calculated.

One disadvantage to consider χ^2 is its dependence on the table dimension and on the total number of observations. The Cramér coefficient, given by

$$\sqrt{\frac{\chi^2/n}{\min(r-1, c-1)}} \, ,$$

is a measure of association in $r \times c$ contingency tables based on χ^2 that is independent of r, c and the total number of observations (n) allowing then comparisons between different tables [1].

Let $C_S(d)$ denote the value of the Cramér coefficient on the contingency table of the variable pairs (*codon, S*) spaced by d genomic symbols S in the total coding sequenced genome. In order to explore the codon bias in a genomic scale and benefit Anaconda outputs, we calculate the levels of the association $C_S(d)$ when $S \in \{codon, nucleotide\}$ and for several values of d.

2.1 DNA Data

For this study we have downloaded the total coding sequences of twelve different species (Table 1) from the RNA files at the National Center for Biotechnology Information ftp site (ftp://ftp.ncbi.nih.gov/genomes/) drawing the complete sets of *Gnomon ab initio* predictions.

2.2 Experimental Procedure

After the acquisition of gene sequences of each specie, we used *Anaconda* tools to convert the data into the contingency tables of two variables mentioned above

Table 1 List of DNA build references for the twelve species considered in the present study

Species	Reference
Homo sapiens (human)	Build 37.1
Pan troglodytes (chimpanzee)	Build 2.1
Rattus norvegicus (brown rat)	Build 4.1
Ornithorhynchus anatinus (platypus)	Build 1.1
Gallus gallus (chicken)	Build 2.1
Apis mellifera (honey bee)	Build 4.1
Vitis vinifera (grape vine)	Build 1.1
Arabidopsis thaliana (thale cress)	AGI 7.2
Saccharomyces cerevisiae str.S228C (budding yeast)	SGD 1
Schizosaccharomyces pombe (fission yeast)	Build 1.1
Escherichia coli str.K12 substr.MG1655 (bacterium)	NC000913
Salmonella typhi (bacterium)	NC003198

selecting $d \in D = \{0,1,2,3,4,5\}$. All tables were obtained excluding the first and the last codon of each gene and with a gene control available in *Anaconda* (test three multiple nucleotide, codon stop, without stop intermediate codon, test start codon, without letters N).

For each specie, we computed $C_S(d)$, $S \in \{codon, nucleotide\}$, and created two vectors (codon bias profiles): one with the levels of the association between codons, $[C_{codon}(d)]_{d \in D}$ and another with the levels of the association between codons and nucleotides, $[C_{nucleotide}(d)]_{d \in D}$. With these profiles we constructed dendrograms to discover relationships between the twelve species considered.

3 Results

Since Pearson chi-squared statistic led to the rejection of null hypothesis of independence (p-value< 0.00001) for all contingency tables generated, we investigated the levels of associations $C_S(d)$, with $S \in \{codon, nucleotide\}$ and $d \in D$. In Figure 1 are illustrated the observed codon bias profiles for each specie.

All the observed levels of association are rather low. In general, the values of $C_{codon}(d)$ decrease when the distance d increases observing greater gap from $d = 0$ to $d = 1$. Nevertheless, as biologically expected [3], the levels of association between contiguous codons are stronger than between codon pairs spaced by $k \geq 1$ codons. For the values $C_{nucleotide}(d)$, it seems there are no global regularities. All species except *A. mellifera* present the highest degree of association between a codon and the first nucleotide of the following one.

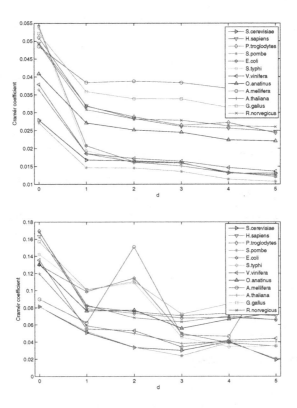

Fig. 1 Observed values of $C_{codon}(d)$ -top- and $C_{nucleotide}(d)$ -bottom- for the twelve species studied

From Figure 1 the codon bias profiles $[C_S(d)]_{d \in D}$ seem to identify each specie, and so may be used as a genomic signature, allowing then the comparison of species. Dendrograms using the set of profiles $[C_{codon}(d)]_{d \in D}$ and $[C_{nucleotide}(d)]_{d \in D}$ of the twelve species were built with complete linkage clustering and Euclidean distance. The profiles $[C_{codon}(d)]_{d \in D}$ allowed to obtain a better dendrogram which can be interpreted as a kind of phylogenetic tree (see Figure 2). It shows hierarchical clusters: bacterias (*S. typhi*, *E. coli*), yeasts (*S. pombe*, *S. cerevisiae*), plants (*A. thaliana*, *V. vinifera*), and animals (*H. sapiens*, *P. troglodytes*, *R. norvegic*, *O. anatinus*, *G. gallus*, *A. mellifera*). For animal cluster, we still distinguish a mammalia cluster.

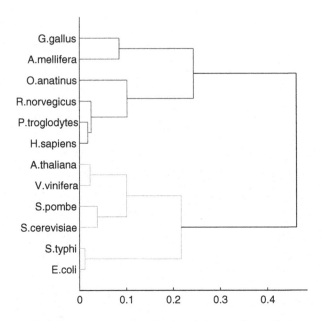

Fig. 2 Phylogenetic tree for the twelve species studied

4   Conclusions and Future Work

The number of complete sequenced genomes is growing up. However, traditional sequence alignment algorithms are a computationally demanding task. In our work a new methodology based on Crámer coefficient is proposed. With the association values of limited number of different codon contexts, we construct a short vector with dimension 5 (codon bias profile).

The low levels of association in all considered contexts of the studied species suggest that the codon bias, when measured in a global manner, represents a small fraction of the set of all features that define the rules in the coding sequences. However, these low levels seem to have a biological interpretation and be sufficiently distinct to create a phylogenetic tree. The dendrograms for the considered species are in accordance with the some expected similarities. Some groups can be identified: animals, plants, fungi, bacteria. All these facts suggest that an analysis of the Crámer coefficient can shed new light into aspects of the codon bias and can help to distinguish species. However, the obtained dendrogram does not separate prokaryotes from eukaryotes.

As future work, we plan to extend this study with more species and to analyze the influence of the dimension of the codon bias profile. Moreover, it is our intention to apply others statistical methodologies to do an automatic taxonomic classification of species using association profiles of each specie.

References

1. Agresti, A.: Categorical Data Analysis. Wiley, Chichester (2002)
2. Berg, O.G., Silva, P.J.N.: Codon bias in *escherichia coli*: The influence of codon context on mutation and selection. Nucleic Acids Research 25, 1397–1404 (1997)
3. Boycheva, S., Chkodrov, G., Ivanov, I.: Codon pairs in genome of *Escherichia coli*. Bioinformatics 19, 987–998 (2002)
4. Carlini, D.B., Stephan, W.: In vivo introduction of unpreferred synonymous codons into the *drosophila* ADH gene results in reduced levels of ADH protein. Genetics 163, 239–243 (2003)
5. Fedorov, A., Saxonov, S., Gilbert, W.: Regularities of context-dependent codon bias in eukaryotic genes. Nucleic Acids Research 30(5), 1192–1197 (2002)
6. Hodge, T., Cope, M.J.T.V.: A myosin family tree. Journal of Cell Science 113, 3353–3354 (2000)
7. Moura, G., Pinheiro, M., Arrais, J., Gomes, A., Carreto, L., Freitas, A., Oliveira, J., Santos, M.: Large scale comparative codon-pair context analysis unveils general rules that fine-tune evolution of mrna primary structure. PLoS ONE 2(9), e847 (2007)
8. Moura, G., Pinheiro, M., Silva, R.M., Miranda, I.M., Afreixo, V.M.A., Dias, G.G., Freitas, A., Oliveira, J.L., Santos, M.: Comparative context analysis of codon pairs on an ORFeome scale. Genome Biology 28(14), R28(14) (2005)
9. Pinheiro, M., Afreixo, V., Moura, G., Freitas, A., Santos, M.A., Oliveira, J.L.: Statistical, computational and visualization methodologies to unveil gene primary structure features. Methods of Information in Medicine 45, 163–168 (2006)
10. Sivaraman, K., Seshasayee, A., Tarwater, P.M., Cole, A.M.: Codon choice in genes depends on flanking sequence informationimplications for theoretical reverse translation. Nucleic Acids Research 36(3), e16 (2008)
11. Wang, F.-P., Li, H.: Codon-pair usage and genome evolution. Gene. 433, 8–15 (2009)
12. Wright, F.: The effective number of codons used in a gene. Gene. 87(1), 23–29 (1990)

An Application for Studying Tandem Repeats in Orthologous Genes

José Paulo Lousado, José Luis Oliveira, Gabriela Moura, and Manuel A.S. Santos

Abstract. Several authors have being suggesting that the occurrence of amino acids repeats in some genes are implied in human diseases. We aim to investigate if these repetitions, which one can observe in humans, also exist in orthologous genes of several organisms, and how they have evolved along time. However, for this kind of study, it is necessary to use different databases and computation methods that are not integrated in any specific tool. In this paper, we present a bioinformatics application, supported by Web Services, that allows to conduct comparative analysis studies for various organisms, along the evolutionary chain.

Keywords: Data integration, Web Services, tandem repeats, codon repetitions.

1 Introduction

The analysis of amino acids sequences of eukaryotic organisms, as well as their evolution over time, has been a highly studied area from the point of view of the evolutionary chain. Several studies [1-4] showed the relationship that exists between some human genes and various illnesses, such as cancer [5], neurodegenerative disorders, and some others [6-7]. Many other studies, which are directed towards the analysis of the human genome, can also be found in the literature. They focus on certain parts of the genomes which have been of the utmost importance for the survival of the human species [8-11]. This refers

José Paulo Lousado
Centro de Estudos em Educação, Tecnologias e Saúde, ESTGL,
Instituto Politécnico de Viseu, Campus Politécnico de Viseu, 3504-510 Viseu, Portugal

José Luis Oliveira
IEETA, University of Aveiro, Campus Universitário de Santiago,
3810-193 Aveiro, Portugal

Gabriela Moura · Manuel A.S. Santos
CESAM, University of Aveiro, Campus Universitário de Santiago,
3810-193 Aveiro, Portugal
e-mail: {lousado, jlo}@ua.pt

M.P. Rocha et al. (Eds.): IWPACBB 2010, AISC 74, pp. 109–115, 2010.
springerlink.com

specifically to the repetitions of certain codons and/or amino acids, which have received the deserved attention of biologists and health researchers so as to be able to predict possible diseases and study specific treatments, namely patient oriented medication [12-15].

In [16], Pearson identifies a set of genes with repetitions which are related to various diseases. Knowing the extent to which these repetitions are present in other organisms which are lower in the evolutionary chain is the main question. The answer to this question may have implications at various levels, particularly in terms of prevention and prophylaxis of certain diseases, as well as the potential spread to descendants of genetic diseases, derived from these replicates.

As a follow up to a previously published paper [17], and given the need to expand the work to other human genes, involving other organisms as well, a data integration computer application was developed, with the aim of facilitating the extraction and comparison of human genes which cause diseases with the respective orthologous genes of different organisms so as to analyse the evolution that has occurred from species to species. Thus, this study was conducted so as to determine whether existing repetitions that cause diseases in humans have propagated to less evolved or more evolved organisms, and how that propagation occurred, with random loss or with an increase in the number of repetitions.

The baseline for the study for which the application was developed was the KEGG database [18]. OMIM [19] was also used for search diseases related to gene repetition.

The data in the KEGG [18] database is separated by organisms and it can be accessed remotely. Querying and relating data of orthologous genes with repetitions are extremely time consuming tasks, exacerbate by the fact that there is no integrated tool that specifically performs these comparisons in an automated manner for a large number of organisms. Creating a bioinformatics application to perform these tasks in an autonomous and integrated manner is, therefore, crucial to obtaining timely data and respective results.

Special relevance is given in this article to the ease of obtaining orthologous genes in a more versatile format so that the data may be saved locally in text files for later offline use. The following goal was the automatic analysis of the amino acid repetitions of the various genes involved.

The multi-window functionality introduced in the application is crucial since the user does not lose data from query to query being able to have up to ten completely independent windows of each of the options. Even if windows are closed, they are easily accessible from the drop-down menu.

2 Methods

For this approach we have used web services that are already available in several biological database, such as KEGG. This web-based technology facilitates data gathering, through a real-time programmatic access. A software application can retrieve information on demand, as it is needed, avoiding downloading a bunch of data, typically through file transfer, just to extract a small sample.

In order to carry out this work a standalone application was developed, following a specific workflow (Fig. 1). It iteratively constructs a relationship between the genes of various organisms using genes that have been previously identified as those which are implicated in diseases, containing codon/amino acid repetitions. Following the work presented by Jones and Pevzner [4], we assume the default value 10 as the minimum representative number of codon and/or amino acid repetitions.

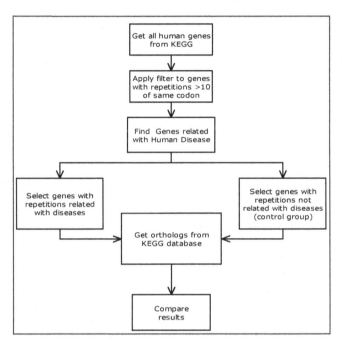

Fig. 1 Data Integration Workflow

The amino acid and codon date are extracted from KEGG reference database. From these files, the application isolates the genes that have replicated codons in sequences, at least with the predefined size threshold.

Once the genes and respective replicated amino acids are identified, a new phase is applied to determine the genes related to diseases. To isolate gene-diseases associations that are caused by tandem repeats, we use OMIM database, since it groups this information in a coherent and effective manner [20].

After that phase, the genes that possess repetitions responsible for diseases are isolated from the remaining genes, which are not directly related to diseases. The purpose of this separation is to create a control group of genes to validate the study. At the end, the results obtained from the test group will be compared with the results from the control group.

From that point, it begins the process of comparing orthologous genes from human to each previously selected organism. A database of human genes and the respective orthologous genes of the various organisms is created.

The application was developed using the .NET platform integrated with Office Web Components.

The use of the KEGG web services, including its integration with the .NET platform requires some modifications in the default parameters to avoid timeouts and transfer breaks (Fig. 2).

```
kegg.KEGGPortTypeClient k;
System.ServiceModel.BasicHttpBinding binding;
binding = DirectCast(k.Endpoint.Binding,
                     System.ServiceModel.BasicHttpBinding);
binding.MaxReceivedMessageSize = 50000000;
binding.MaxBufferSize = 50000000;
binding.ReaderQuotas = System.Xml.XmlDictionaryReaderQuotas.Max;
binding.SendTimeout = TimeSpan.FromMinutes(2);
```

Fig. 2 Changes applied on KEGG Web Services parameters

Thus problems related to the data buffer memory capacity via http and timeout errors can be overcome. The errors of passing data via http occur when the data buffer is exhausted. This value is defined as 64Kb by default, which is clearly insufficient since there are genes whose associated information greatly exceeds this value. Setting this parameter with a value close to 50Mb practically guarantees the collection of data without errors. To avoid situations in which the system may use time-consuming connection resources, the timeout parameter was set at 2 minutes.

The Framework is made up of two modules: "Data Retrieve" and "Orthologous Search". The first module is an exploration tool isolated from the database (back office). This allows the system to return the respective orthologous at the desired depth by merely introducing the gene or KO, e.g. hsa:367 or ko:K08557. The information collected is then displayed in the application's frames. The respective orthologous genes are displayed on the left-hand list as well as the information on diseases related to that gene and pathways, if such is the case. The user may then access the orthologous by simply selecting the respective gene from the list on the left. The data that are being viewed in the amino acid and nucleotide frames are held in memory by default. The user may save the data by simply selecting the respective option (Fig. 3). By accessing the list of diseases, the window shows all of the available information including references about each disease.

The "Orthologous Search" module is essentially an application of batch processing, that is, once the user creates the list of genes to be analysed and the list of organisms to be compared, the system will submit data to the KEGG database, automatically and iteratively. It extracts the information and saves the respective file in the folder that has been previously selected for that purpose.

As the data are being processed, a list of orthologous genes in the selected organisms is created. The order in the final list depends on the degree of similarity against the original sequence. If the orthologous genes in one or more organisms do not appear, the *Limit* field may be adjusted to a higher value. For example, *Limit=60* and *Offset=20* will return orthologous genes that are found between position 20 and 80 in the KEGG database for the orthologous genes being studied.

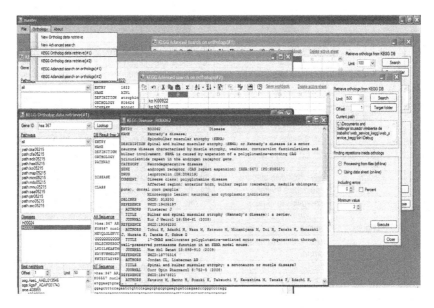

Fig. 3 Application Interface With Multiple Windows

The tool also incorporates other functionalities, including the search of repetitions of amino acids with or without the admission of errors by the orthologous of one or more genes placed in the first column. Thus, we can resort to the use of the local files previously saved in Batch Processing, or opt for using a connection to the Web. In this case performance may go down, but on the other hand more up-to-date results may be obtained.

After processing, several spreadsheets are created with the results of the analysis on each of the respective orthologous. These spreadsheets can be processed locally as a single file in XLS format.

3 Discussion

To test the application, we have conducted an analysis in accordance with the previously presented protocol (Figure 1), comparing human genes under the referred conditions with the respective orthologous genes of the organisms presented in Table 1. Obtaining the results was almost immediate for the local source (offline), and it takes only a few seconds, depending on bandwidth, when we use directly the web server as a source of data (online). Integrating scattered data, whether via the Web (several sources), whether by post-processing (offline files), the developed tool becomes a crucial ally for researchers, mainly in situations of massive data extraction.

Table 1 Organisms list and genealogic classification

vertebrate	mammals	viviparous	Bos taurus	ftp://ftp.genome.jp/pub/kegg/genes/organisms/bta
			Canis familiaris	ftp://ftp.genome.jp/pub/kegg/genes/organisms/cfa
			Homo sapiens	ftp://ftp.genome.jp/pub/kegg/genes/organisms/hsa
			Mus musculus	ftp://ftp.genome.jp/pub/kegg/genes/organisms/mmu
			Pan troglodytes	ftp://ftp.genome.jp/pub/kegg/genes/organisms/ptr
		marsupial	Monodelphis domestica	ftp://ftp.genome.jp/pub/kegg/genes/organisms/mdo
		oviparous	Ornithorhynchus anatinus	ftp://ftp.genome.jp/pub/kegg/genes/organisms/oaa
	bird		Gallus gallus	ftp://ftp.genome.jp/pub/kegg/genes/organisms/gga
	fish		Danio rerio	ftp://ftp.genome.jp/pub/kegg/genes/organisms/dre
invertebrate	Insect		Drosophila melanogaster	ftp://ftp.genome.jp/pub/kegg/genes/organisms/dme
	worm		Caenorhabditis elegans	ftp://ftp.genome.jp/pub/kegg/genes/organisms/cel
	plant		Arabidopsis thaliana	ftp://ftp.genome.jp/pub/kegg/genes/organisms/ath
	fungus		Aspergillus fumigatus	ftp://ftp.genome.jp/pub/kegg/genes/organisms/afm
			Kluyveromyces lactis	ftp://ftp.genome.jp/pub/kegg/genes/organisms/kla
			Saccharomyces cerevisiae	ftp://ftp.genome.jp/pub/kegg/genes/organisms/sce
			Schizosaccharomyces pombe	ftp://ftp.genome.jp/pub/kegg/genes/organisms/spo
	protozoan		Plasmodium falciparum	ftp://ftp.genome.jp/pub/kegg/genes/organisms/pfa
	bacteria		Clostridium perfringens	ftp://ftp.genome.jp/pub/kegg/genes/organisms/cpe
			Mycobacterium tuberculosis	ftp://ftp.genome.jp/pub/kegg/genes/organisms/mtu

4 Conclusion

Codon repeats in DNA are contiguous, and some time approximate, copies of a pattern of trinucleotides. These repeats have been associated to specific human diseases, and may play a variety of regulatory and evolutionary roles.

In this paper we presented a computation application that simplifies the study of genes with this type of patterns, along the evolutionary chain. For that the software extracts orthologous genes from public resources and performs a comparative analysis that allows showing how repeats have evolved along the time for the specie other study.

Acknowledgements

J. P. Lousado has funded by the Instituto Politécnico de Viseu under the PROFAD Programme.

References

1. George, R.A., Liu, J.Y., Feng, L.L., Bryson-Richardson, R.J., Fatkin, D., Wouters, M.A.: Analysis of protein sequence and interaction data for candidate disease gene prediction. Nucl. Acids Res. 34, e130 (2006)
2. Sher Ali, S.A., Ehtesham, N.Z., Azfer, M.A., Homkar, U., Rajesh Gopal, S.E.H.: Analysis of the evolutionarily conserved repeat motifs in the genome of the highly endangered central Indian swamp deer Cervus duvauceli branderi. GENE. 223, 361–367 (1998)
3. Fu, Z., Jiang, T.: Clustering of main orthologs for multiple genomes. J. Bioinform. Comput. Biol. 6, 573–584 (2008)

4. Jones, N.C., Pevzner, P.A.: Comparative genomics reveals unusually long motifs in mammalian genomes. Bioinformatics 22, e236–e242 (2006)

5. Lab, P.: Repeat Disease Database (2009), http://www.cepearsonlab.com/rdd.php

6. Brameier, M., Wiuf, C.: Ab initio identification of human microRNAs based on structure motifs. BMC Bioinformatics 8, 478 (2007)

7. Bowen, T.A., Guy, C.A.A., Cardno, A.G.A.B., Vincent, J.B.C., Kennedy, J.L.C., Jones, L.A.A., Gray, M.A., Sanders, R.D.A., McCarthy, G.A., Murphy, K.C.A., Owen, M.J.A.B., O'Donovan, M.C.A.: Repeat sizes at CAG/CTG loci CTG18.1, ERDA1 and TGC13-7a in schizophrenia. Psychiatric Genetics 10, 33–37 (2000)

8. Ferro, P., Catalano, M.G., Dell'Eva, R., Fortunati, N., Pfeffer, U.: The androgen receptor CAG repeat: a modifier of carcinogenesis? Molecular and Cellular Endocrinology 193, 109–120 (2002)

9. Pestova, T.V., Hellen, C.U., Wimmer, E.: A conserved AUG triplet in the 5' nontranslated region of poliovirus can function as an initiation codon in vitro and in vivo. Virology 204, 729–737 (1994)

10. Freed, K.A., Cooper, D.W., Brennecke, S.P., Moses, E.K.: Detection of CAG repeats in pre-eclampsia/eclampsia using the repeat expansion detection method. Mol. Hum. Reprod. 11, 481–487 (2005)

11. Herishanu, Y.O., Parvari, R., Pollack, Y., Shelef, I., Marom, B., Martino, T., Cannella, M., Squitieri, F.: Huntington disease in subjects from an Israeli Karaite community carrying alleles of intermediate and expanded CAG repeats in the HTT gene: Huntington disease or phenocopy? Journal of the Neurological Sciences 277, 143–146 (2009)

12. Bogaerts, V., Theuns, J., van Broeckhoven, C.: Genetic findings in Parkinson's disease and translation into treatment: a leading role for mitochondria? Genes. Brain. Behav. 7, 129–151 (2008)

13. Mena, M.A., Rodriguez-Navarro, J.A., Ros, R., de Yebenes, J.G.: On the pathogenesis and neuroprotective treatment of Parkinson disease: what have we learned from the genetic forms of this disease? Curr. Med. Chem. 15, 2305–2320 (2008)

14. Tarini, B.A., Singer, D., Clark, S.J., Davis, M.M.: Parents' Interest in Predictive Genetic Testing for Their Children When a Disease Has No Treatment. Pediatrics (2009)

15. Hsueh, W.: Genetic discoveries as the basis of personalized therapy: rosiglitazone treatment of Alzheimer's disease. Pharmacogenomics J. 6, 222–224 (2006)

16. Pearson, C.E., Edumura, N.K., Cleary, J.D.: instability: mechanisms of dynamic mutations. Nat. Rev. Genet. 6, 729–742 (2005)

17. Lousado, J., Oliveira, J., Moura, G., Santos, M.: Analysing the Evolution of Repetitive Strands in Genomes. In: Omatu, S., Rocha, M.P., Bravo, J., Fernández, F., Corchado, E., Bustillo, A., Corchado, J.M. (eds.) IWANN 2009. LNCS, vol. 5518, pp. 1047–1054. Springer, Heidelberg (2009)

18. KEGG: Kyoto Encyclopedia of Genes and Genomes. Kanehisa Laboratories, http://www.kegg.com

19. OMIM, Online Mendelian Inheritance in Man, http://www.ncbi.nlm.nih.gov/omim/

20. Hamosh, A., Scott, A., Amberger, J., Bocchini, C., McKusick, V.: Online Mendelian Inheritance in Man (OMIM), a knowledgebase of human genes and genetic disorders. Nucleic Acids Research 33, D514 (2005)

21. Kanehisa, M., Goto, S., Furumichi, M., Tanabe, M., Hirakawa, M.: KEGG for representation and analysis of molecular networks involving diseases and drugs. Nucl. Acids Res. 38, D355–D360 (2010)

Accurate Selection of Models of Protein Evolution

Mateus Patricio, Federico Abascal, Rafael Zardoya, and David Posada

Abstract. We carried out computer simulations to assess the accuracy of the program ProtTest in recovering best-fit empirical models of amino acid replacement. We were able to show that regardless of the selection criteria used, the simulated model or a very close one was identified most of the time. In addition, the estimates of the different model parameters were very accurate. Our results suggest that protein model selection works reasonably well.

1 Introduction

The use of different models of nucleotide substitution or amino acid replacement can change the outcome of the phylogenetic analysis [1]. Apart from the different effects on parameter estimation and hypothesis testing, in general, phylogenetic methods may be less accurate –recover more often a wrong tree–, or may be inconsistent –converge to a wrong tree with increased amounts of data– when the assumed model of evolution is wrong. Conveniently, best-fit models of molecular evolution can be selected for the data at hand using sound statistical techniques, like hierarchical likelihood ratio tests, information criteria, Bayesian methods or performance-based approaches [2]. Several of these strategies have been implemented in different computer programs like ModelTest [3] and jModelTest [4] for DNA sequences, and ProtTest [5] for proteins. The performance of ModelTest was examined in detail by Posada and Crandall [6], who showed that model selection was very accurate in the case of nucleotides. However, we do not know whether this is the case for protein sequences. In order to answer this question we have used computer simulations to evaluate the ability of ProtTest in recovering the simulated model of amino acid replacement.

Mateus Patricio · David Posada
Departamento de Bioquímica, Genética, e Inmunología, Universidad de Vigo, Vigo, Spain

Federico Abascal · Rafael Zardoya
Departamento de Biodiversidad y Biología Evolutiva,
Museo Nacional de Ciencias Naturales, Madrid, Spain

M.P. Rocha et al. (Eds.): IWPACBB 2010, AISC 74, pp. 117–121, 2010.
springerlink.com   © Springer-Verlag Berlin Heidelberg 2010

2 Materials and Methods

To simulate the protein alignments, we downloaded from http://www.atgc-montpellier.fr/phyml/datasets.php the set of trees that were used to evaluate the performance of the Phyml algorithm [7] for phylogenetic estimation. This set contains 5000 simulated non-clock trees comprising 40-taxon each, and includes a variety of topologies and tree heights. For each one of these trees, we generated a 40-sequence alignment with 1000 amino acids using Seq-Gen [8], under the JTT [9] matrix of amino acid substitution, with rate variation among sites (number of gamma rate categories = 4; alpha gamma shape = 0.5; proportion of invariable sites = 0.2). This model, whose acronym is JTT+I+G, will be referenced hereafter as the *true* model. For each alignment, the best-fit model of amino acid replacement was selected with ProtTest among 112 candidate models using the uncorrected and corrected Akaike Information Criterion (AIC and AICc) and the Bayesian Information Criterion (BIC), including model-averaged estimates and parameter importances (see [2]). For the AICc and BIC, sample size was calculated as the number of sites in the alignment. The calculation of the likelihood score and maximum likelihood (ML) tree for each model is very time-consuming, and in order to speed up the whole process all the program executions were carried out using distributed computing techniques. After running ProtTest, we counted the amount of times the generating model was correctly identified and calculated the mean squared error (MSE) of the parameter estimates.

3 Results

ProtTest correctly identified the true model (JTT+I+G; ~90%) or very closely related model (JTT+G; 10%) most of the time (Table 1). In general, the best-fit model was selected with little uncertainty and therefore receiving large weights (Table 2).

Table 1 Number of times each model was identified as the best-fit model (percentages in parentheses)

Model	JTT+G	JTT+I+G	JTT+G+F	JTT+I
AIC	523 (0.10)	4466 (0.90)	1 (0)	6 (0.001)
AICc	553 (0.11)	4437 (0.90)	6 (0.01)	–
BIC	1034 (0.21)	3956 (0.79)	–	6 (0.001)

Table 2 Average/median model weights for the different models according to the three model selection criteria

Model	True	Best	2nd Best	JTT+G
AIC	0.90	0.90	0.31	0.31
AICc	0.89	0.89	0.33	0.33
BIC	0.79	0.79	0.52	0.52

The estimates of the *alpha* and *p-inv* parameters were very accurate, regardless of whether they corresponded to the best-fit model, to the true model, or to model-averaged estimates (Fig. 1 and Tables 3 and 4).

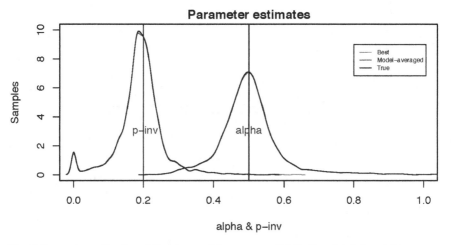

Fig. 1 Parameter estimates of the shape of the gamma distribution (*alpha*) and the proportion of invariable sites (*p-inv*) for the best-fit model, true model and model-averaged estimates (the three lines are almost perfectly overlapping) under the AIC framework.. Vertical lines correspond to the true simulated values

Table 3 Mean squared error (MSE), mean (Mean) and standard deviation (Sd) of the estimated shape of the gamma distribution (*alpha*)

	MSE	Mean	Sd
Best-fit model	0.016	0.508	0.115
True model	0.013	0.508	0.115
Model-averaged	0.013	0.508	0.115

Table 4 Mean squared error (MSE), mean (Mean) and standard deviation (Sd) for the estimated proportion of invariable sites (*p-inv*)

	MSE	Mean	Sd
Best-fit model	0.004	0.190	0.065
True model	0.004	0.190	0.064
Model-averaged	0.004	0.190	0.064

The importance of the parameters (Fig. 2) shows that the most important parameter is the joint contribution made by the proportion of invariable sites plus the shape of the gamma distribution (*p-inv + alpha*), which is the expected since the true model includes both (JTT+I+G).

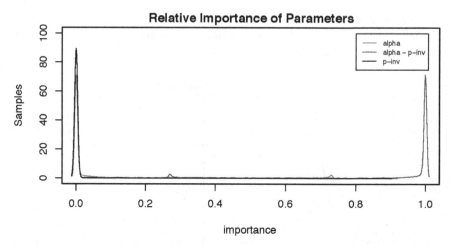

Fig. 2 Relative importance of parameters. *alpha*: Mean = 0.09785 and Sd = 0.20990. *p-inv*: Mean = 0.00176, and Sd = 0.02630. *alpha+p-inv*: Mean = 0.90037, Median: 1, Var = 0.04595 and Sd = 0.21436.

4 Discussion

In our simulations ProtTest was able to identify, regardless the selection criteria used, the true generating model (JTT+I+G) or a closely related one (JTT+G) most of the time. Given that the only difference between these two models is that the former assumes a proportion of invariable sites (here *p-inv* = 0.2), while in the latter this proportion is, by definition, zero, it is not surprising that in some cases JTT+G was preferred over JTT+I+G. The estimation of the different parameters was also very accurate under the thee model selection frameworks explored. Here, the estimates provided by the best-fit and true models or using model averaging, were very similar, which was expected given that the best-fit model usually corresponded with the true model and received large weights. Also, the parameter importances reflected very well the need for taking into account both a proportion of invariable sites and rate variation among the variable sites, in agreement with the true model.

5 Conclusions

Protein model selection using ProtTest seems to work very well. The AIC and BIC frameworks provides accurate tools for the study of the process of molecular evolution at the protein level. Further simulations should explore other generating models and quantify the effect of protein model selection on phylogenetic accuracy.

References

1. Sullivan, J., Joyce, P.: Model selection in phylogenetics. Annual Review of Ecology. Evolution and Systematics 36, 445–466 (2005)
2. Posada, D., Buckley, T.R.: Model selection and model averaging in phylogenetics: advantages of Akaike information criterion and Bayesian approaches over likelihood ratio tests. Syst. Biol. 53(5), 793–808 (2004)
3. Posada, D., Crandall, K.A.: Modeltest: testing the model of DNA substitution. Bioinformatics 14(9), 817–818 (1998)
4. Posada, D.: jModelTest: phylogenetic model averaging. Mol. Biol. Evol. 25(7), 1253–1256 (2008)
5. Abascal, F., Zardoya, R., Posada, D.: ProtTest: selection of best-fit models of protein evolution. Bioinformatics 21(9), 2104–2105 (2005)
6. Posada, D., Crandall, K.A.: Selecting the best-fit model of nucleotide substitution. Systematic Biology 50, 580–601 (2001)
7. Guindon, S., Gascuel, O.: A simple, fast, and accurate algorithm to estimate large phylogenies by maximum likelihood. Syst. Biol. 52(5), 696–704 (2003)
8. Rambaut, A., Grassly, N.C.: Seq-Gen: an application for the Monte Carlo simulation of DNA sequence evolution along phylogenetic trees. Comput. Appl. Biosci. 13(3), 235–238 (1997)
9. Jones, D.T., Taylor, W.R., Thornton, J.M.: The rapid generation of mutation data matrices from protein sequences. Comput. Appl. Biosci. 8(3), 275–282 (1992)

Scalable Phylogenetics through Input Preprocessing

Roberto Blanco, Elvira Mayordomo,
Esther Montes, Rafael Mayo, and Angelines Alberto

Abstract. Phylogenetic reconstruction is one of the fundamental problems in computational biology. The combinatorial explosion of the state space and the complexity of mathematical models impose practical limits on workable problem sizes. In this article we explore the scalability of popular algorithms under real datasets as problem dimensions grow. We furthermore develop an efficient preclassification and partitioning strategy based on guide trees, which are used to intently define an evolutionary hierarchy of groups of related data, and to determine membership of individual data to their corresponding subproblems. Finally, we apply this method to efficiently calculate exhaustive phylogenies of human mitochondrial DNA according to phylogeographic criteria.

1 Motivation

The organization of living organisms (extant or not) into a "tree of life", as conceived by Darwin, is the purpose of the modern discipline of phylogenetics [4]. Continuous advances in sequencing technologies have supported an exponential growth in publicly available biological sequences over the last quarter century. This abundance offers extraordinary potential to shed light into the inner workings of evolution.

Phylogenetic techniques infer trees, or generalizations thereof, from multiple sequence alignments according to a certain optimality criterion. This mathematical

Roberto Blanco · Elvira Mayordomo
Departamento de Informática e Ingeniería de Sistemas (DIIS)/Instituto de Investigación en Ingeniería de Aragón (I3A), Universidad de Zaragoza, María de Luna 1,
50018 Zaragoza, Spain
e-mail: {robertob,elvira}@unizar.es

Angelines Alberto · Rafael Mayo · Esther Montes
Centro de Investigaciones Energéticas, Medioambientales y Tecnológicas (CIEMAT),
Avenida Complutense 22, 28040 Madrid, Spain
e-mail: {angelines.alberto,rafael.mayo,esther.montes}@ciemat.es

M.P. Rocha et al. (Eds.): IWPACBB 2010, AISC 74, pp. 123–130, 2010.
springerlink.com   © Springer-Verlag Berlin Heidelberg 2010

scoring scheme acts as a computational surrogate of the true biological objective: to recover the true evolutionary relationships between living organisms, as represented by (usually) aligned, meaningful sequences.

Nevertheless, this is a very challenging problem. Not only are there no efficient methods to calculate an optimal phylogeny given an alignment and a tree scoring function, but the preservation of the optimality of a solution while adding new data to it has been proven to be NP-hard for even the elementary parsimony criterion [3]. Therefore, standard practice depends on heuristic methods, which still remain very costly.

From an information-rich perspective, the main defect of conventional methods is that they are completely blind, whereas it would be desirable to provide hints and facts that may constrain and help to simplify problem solution. In this paper, we resort to evolutionary hypotheses as classifiers and definers of smaller, simpler subproblems, and illustrate the biological and computational benefits of such a methodology.

2 Methods

In this section we introduce the working principles of our technique, providing the experimental grounds for it as well as its theoretical basis. The illustration of the practice and benefits of this framework is deferred to the next section.

2.1 Stand-Alone Algorithms

There exist many phylogenetic reconstruction methods, but their principal classification regards whether they treat statistical quality evaluation as an additional step or as part of the process itself [6]. This assessment is of the utmost importance because of the infeasibility of exact algorithms for even small datasets. Both approaches have strengths and weaknesses, and thus method selection is a very important choice, not least due to the very poor scaling of most techniques.

To assess the severity of these decisions, we have evaluated the relative performance and scalability of the main families of methods; our results are summarized in Fig. 1. Problem dimensions have been chosen to scale up to moderately sized, real datasets, with which we will deal in Sect. 3. PHYLIP and MrBayes [5, 8] have been used for traditional and Bayesian methods, respectively, due to their availability and generality.

All three traditional methods are seen to follow similar trends, though parsimony is more abrupt in its progression. Generally speaking, more thorough methods incur in significantly higher costs, quickly becoming impractical. It should be noted that these times must be multiplied by the desired number of bootstrap samples for the analysis, which can nevertheless be executed in parallel.

The raw computational needs of Bayesian methods are more difficult to estimate due to their simulational nature, as is their parallelization. Cost ultimately depends on both problem dimensions, number of iterations (and its relation to model convergence and stop conditions), and possibly number of processors. For the sake of measurement we have fixed an appropriate, catch-all number of iterations and assume approximately linear speed-up for the execution setup [1]. Time growth is comparable, though steeper and orders of magnitude above likelihood search, which together with indivisibility make these methods impracticable for large datasets.

Generally speaking, it can be concluded that, since problem complexity is always superlinear, partitioning offers very significant improvements and allows agile production of cleaner results; therefore, it should be exploited whenever possible. These gains can be invested in the calculation of a higher degree of statistical support or the selection of more sophisticated methods and substitution models.

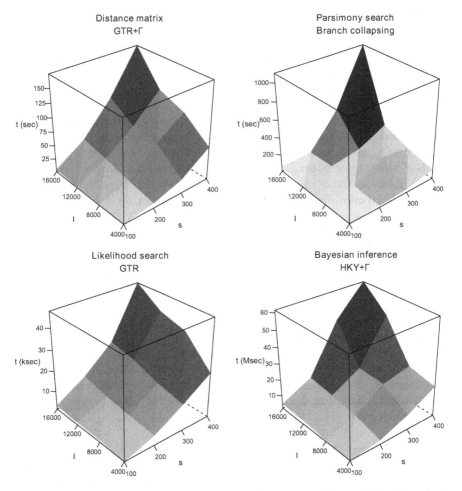

Fig. 1 Performance and scalability of phylogenetic reconstruction methods along both problem dimensions: number of sequences s and sequence length l

2.2   The Supertree Approach

Supertree methods have been conceived to combine, and possibly reconcile, a number of input subtrees according to the underlying relations between them that are manifested through a set of overlapping taxa, which are used to puzzle the trees together [10]. This approach is advantageous in that it is ideally cost-effective and allows comprehensive studies while preserving known results.

If inputs are reasonably structured and compatible, i.e., there are no unsolvable contradictions between the inferred sets of clades, it is possible to produce exact solutions in polynomial time [11]. What we develop here is a further simplification of the compositional burden based on prior structural knowledge about the data.

Our proposal is an extension of the hierarchical classification problem. If a decision tree can be provided that allows simple, recursive categorization of single sequences and that reflects well-supported phylogenetic knowledge, such a "skeleton tree" can be employed to obtain groups of related sequences, thus splitting the data into their corresponding subclades. Each of these can be subsequently computed independently and in parallel. In essence, we use supertrees to reduce problem dimensionality through a classic divide and conquer scheme.

Insofar as we know for certain, or are willing to assume as hypotheses, the relations between the clades defined by the preprocessing hierarchy, composition of partial results into the final tree is limited to substitution of symbolic "clade nodes" by their associated subtrees. The substitution skeleton must, however, accommodate all clades as leaves of the tree, transferring internal categories down dummy leaf branches if needed. The process is straightforward save for this provision.

3   Case Study: Human Mitochondrial DNA

Mitochondrial DNA (mtDNA) is one of the most important evolutionary markers for phylogenetics due to its remarkable features: absence of effective recombination, high mutation rate and ease of sequencing, among others. Its prime metabolic roles also grant it prominence in the study of rare genetic disease [13].

Consequently, comprehensive research on human mtDNA is of great interest, though the very own wealth of available information deters straightforward, manually supervised trees; we have previously addressed the question in [2] and subsequent work. At present we can effectively perform periodic updates to the human mitochondrial phylogeny, though as we endeavor to show there is plenty of room for improvement.

3.1   Structural Properties

For the proposed supertree methods to be applicable, firstly a suitable set of classifiers needs to be identified. In the case of mtDNA, its matrilineal inheritance along with the migrations that scattered the human species around the world gave rise to mitochondrial haplogroups: large population groups related by common descent,

Fig. 2 Phylogeographical history of the major human mitochondrial haplogroups, fitted to MITOMAP's tree skeleton. The star marks the root of the tree. Internal clades that were excluded from the hierarchy are greyed out

to which membership can be ascribed simply by checking a handful of defining polymorphisms.

These groups spread in an arborescent fashion, not unlike a conventional species tree, as depicted in Fig. 2. The study of human haplogroups is a well-founded area of ongoing research with diverse applications [12] and a standard cladistic notation established in [7]. It is therefore perfectly adequate to our purposes as a recipient of phylogenetic knowledge.

3.2 Materials and Methods

For the conduction of our experiments we have prepared a curated database of completely sequenced, human mtDNA sequences, partly based on MITOMAP's query on GenBank. The aligned sequences ($s = 4895$, $l = 16707$) can be assumed to be homogeneous and show no obvious flaws.

Also from the MITOMAP phylogeny [9] we have selected its "Simplified mtDNA lineages" as an adequate guide tree that offers a good level of detail for a first classification. Minor corrections have been made regarding haplogroups B (allowing room for ambiguity in the definition of its distinctive indel) and I (avoiding a reticulation event for which no evidence has been found and promoting it as a direct descendant of N).

Recursive classification has been performed to assign each sequence to its related haplogroup; sequences that fail to match exactly one category have been excluded from the analysis (1.7% of the total, a very small fraction given the low complexity of the skeleton). Haplogroup subtrees have been computed using distance matrices with neighbor joining clustering and bootstrap sampling, and Bayesian inference; both under adequate substitution models. Executions have been distributed across high-performance clusters to exploit the potential parallelism unveiled by our method.

3.3   Results

As a result of the classification we obtain 28 clade subsets with sizes up to $s = 583$ owing to unequal sequencing of population groups around the globe. Due to the progression of computation times with s, a handful of prolific groups clearly dominates total execution times. However, the recursive nature of the tree makes it possible to easily refine copious nodes by simply identifying and adding children

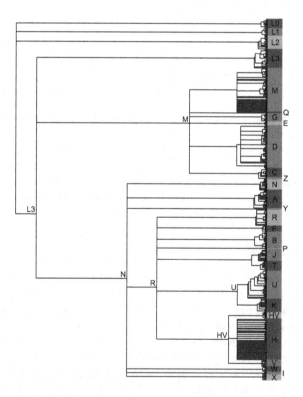

Fig. 3 Human mitochondrial haplogroup supertree. Parent haplogroups have been propagated to the top of their subtrees as leaf haplogroups; they are also labels of their associated clades

subtrees as required; this has indeed been the case of Bayesian inference of groups D, H, M and U, due to technical limitations. In fact, although traditional methods yield satisfactory throughput, Bayesian inference remains only marginally tractable and on the whole depends on more extensive partitioning.

The improvements derived from preprocessing are very remarkable. As a reference, the cost of computing a single distance matrix for the selected dataset with a suitable, reasonably complex evolutionary model is approximately 10^5 sec. By contrast, a single workstation requires the same time to produce reasonably supported (e.g., 100 bootstrap samples per haplogroup) combined phylogenies under the same algorithm and model, therefore amounting roughly to a 100-fold increase in performance; additional gains can be achieved by exploiting the larger number of simpler, less costly tasks.

The phylogenies we obtain are qualitatively better than standard, blind results in that results are clear, and noise and discordance are confined to individual subproblems, hence bounding their potential effect and improving overall quality and robustness. One such phylogeny can be seen in Fig. 3.

4 Discussion and Future Work

We have presented an efficient and effective supertree technique that effectively performs a reduction on s, the costliest dimension of the phylogeny reconstruction problem, through a divide and conquer approach where the penalties of the split and merge operations are negligible. As a result, computational load is substantially lowered, known properties are respected by construction, and the number of completely independent, distributable tasks is increased. These improvements afford more detailed treatment of individual problem instances and feasible resolution of continuously growing inputs.

We also deem it possible to develop reasonably accurate cost prediction, which, despite some measure of input dependence and irregularity, may be of great assistance in selecting methods (and models) in accord with available time and resources. Such estimates may also lead to better scheduling and load balancing in distributed environments.

It has been stressed that solutions are, as a matter of fact, qualitatively better. Moreover, we would like to know how prior knowledge may affect quantitative scores when pitted against atomic methods, as well as the effect and progression of individual algorithms and substitution models, and pinpoint the sources of error, including those that might cause incorrect group ascription.

Finally, considering that reliable phylogenetic knowledge is the source of all classification hierarchies, their appraisal merits further attention. The fit of an alignment and a guide tree, or lack thereof, may evidence a need for (possibly iterative) refinement of one or both parts. Likewise, the capacity to generate restricted datasets and treat them in finer detail could be used to support (sub)haplogroup identification while providing feedback and increasing guide tree resolution.

Acknowledgements. This work was supported in part by the Spanish Ministry of Science and Innovation (MICINN) under Action CAC-2007-52 and Project TIN2008-06582-C03-02. Roberto Blanco was supported by Grant AP2008-03447 from the Spanish Ministry of Education.

References

1. Altekar, G., Dwarkadas, S., Huelsenbeck, J.P., Ronquist, F.: Parallel Metropolis coupled Markov chain Monte Carlo for Bayesian phylogenetic inference. Bioinformatics 20, 407–415 (2004)
2. Blanco, R., Mayordomo, E.: ZARAMIT: a system for the evolutionary study of human mitochondrial DNA. In: Omatu, S., Rocha, M.P., Bravo, J., Fernández, F., Corchado, E., Bustillo, A., Corchado, J.M. (eds.) IWANN 2009. LNCS, vol. 5518, pp. 1139–1142. Springer, Heidelberg (2009)
3. Böckenhauer, H.J., Hromkovič, J., Královič, R., Mömkea, T., Rossmanith, P.: Reoptimization of Steiner trees: changing the terminal set. Theor. Comput. Sci. 410, 3428–3435 (2009)
4. Ciccarelli, F.D., Doerks, T., von Mering, C., Creevey, C.J., Snel, B., Bork, P.: Toward automatic reconstruction of a highly resolved tree of life. Science 311, 1283–1287 (2006)
5. Felsenstein, J.: PHYLIP – Phylogeny Inference Package (version 3.2). Cladistics 5, 164–166 (1989)
6. Holder, M., Lewis, P.O.: Phylogeny estimation: traditional and Bayesian approaches. Nat. Rev. Genet. 4, 275–284 (2003)
7. Richards, M.B., Macaulay, V.A., Bandelt, H.J., Sykes, B.C.: Phylogeography of mitochondrial DNA in western Europe. Ann. Hum. Genet. 62, 241–260 (1998)
8. Ronquist, F., Huelsenbeck, J.P.: MrBayes 3: Bayesian phylogenetic inference under mixed models. Bioinformatics 19, 1572–1574 (2003)
9. Ruiz-Pesini, E., Lott, M.T., Procaccio, V., Poole, J.C., Brandon, M.C., Mishmar, D., Yi, C., Kreuziger, J., Baldi, P., Wallace, D.C.: An enhanced MITOMAP with a global mtDNA mutational phylogeny. Nucleic Acids Res. 35, D823–D828 (2007)
10. Sanderson, M.J., Purvis, A., Henze, C.: Phylogenetic supertrees: assembling the trees of life. Trends Ecol. Evol. 13, 105–109 (1998)
11. Steel, M., Dress, A.W.M., Böcker, S.: Simple but fundamental limitations on supertree and consensus tree methods. Syst. Biol. 49, 363–368 (2000)
12. Torroni, A., Achilli, A., Macaulay, V., Richards, M., Bandelt, H.J.: Harvesting the fruit of the human mtDNA tree. Trends Genet. 22, 339–345 (2006)
13. Wallace, D.C.: A mitochondrial paradigm of metabolic and degenerative diseases, aging, and cancer: a dawn for evolutionary medicine. Annu. Rev. Genet. 39, 359–407 (2005)

The Median of the Distance between Two Leaves in a Phylogenetic Tree

Arnau Mir and Francesc Rosselló

Abstract. We establish a limit formula for the median of the distance between two leaves in a fully resolved unrooted phylogenetic tree with n leaves. More precisely, we prove that this median is equal, in the limit, to $\sqrt{4\ln(2)n}$.

1 Introduction

The definition and study of metrics for the comparison of phylogenetic trees is a classical problem in phylogenetics [5, Ch. 30], motivated, among other applications, by the need to compare alternative phylogenies for a given set of organisms obtained from different datasets or using different methods. Many metrics for the comparison of rooted or unrooted phylogenetic trees on the same set of taxa have been proposed so far. Some of the most popular, and oldest, such metrics are based on the comparison of the vectors of distances between pairs of taxa in the corresponding trees [1, 3, 4, 9, 10, 11]. These metrics aim at the quantification of the rate at which pairs of taxa that are close together in one tree lie at opposite ends in another tree [7]. But, in contrast with other metrics, their statistical properties are mostly unknown.

Steel and Penny [10] computed the mean value of the square of the metric for fully resolved unrooted trees defined through the euclidean distance between their vectors of distances (they called it the *path difference metric*). One of the main ingredients in their work was the explicit computation of the mean value and the variance of the distance d between two leaves in a fully resolved unrooted phylogenetic tree with n leaves, obtaining that

$$\mu(d) = \frac{2^{2(n-2)}}{\binom{2(n-2)}{n-2}} \sim \sqrt{\pi n}, \quad \mathrm{Var}(d) = 4n - 6 - \mu(d) - \mu(d)^2$$

Arnau Mir · Francesc Rosselló
Department of Math & Computer Science, University of Balearic Islands, Crta. Valldemossa, km. 7,5, 07122 Palma de Mca, Spain
e-mail: {arnau.mir,cesc.rossello}@uib.es

M.P. Rocha et al. (Eds.): IWPACBB 2010, AISC 74, pp. 131–135, 2010.
springerlink.com   © Springer-Verlag Berlin Heidelberg 2010

Such values are also of some interest in themselves, because they give figures against which to compare the corresponding values for the distribution of distances between leaves in a reconstructed phylogenetic tree, allowing one to assess how "close to average" is such a tree. Some specific applications of the probabilistic behaviour of parameters related to the shape of phylogenetic trees are discussed in [2].

In this work we continue the statistical analysis of this random variable d, by giving an expression for its median that allows the derivation of a limit formula for it: namely,

$$\text{median}(d) \sim \sqrt{4n \ln 2}.$$

The derivation of a limit formula for the median of the aforementioned squared path difference metric between fully resolved unrooted phylogenetic trees remains an open problem. We hope this result will constitute a first step towards its achievement. We shall report on this application elsewhere.

2 Preliminaries

In this paper, by a *phylogenetic tree* on a set S we mean a *fully resolved* (that is, with all its internal nodes of degree 3) unrooted tree with its leaves bijectively labeled in the set S. Although in practice S may be any set of taxa, to fix ideas we shall always take $S = \{1, \ldots, n\}$, with n the number of leaves of the tree, and we shall use the term *phylogenetic tree with n leaves* to refer to a phylogenetic tree on this set. For simplicity, we shall always identify a leaf of a phylogenetic tree with its label.

Let $\mathcal{T}_n$ be the set of (isomorphism classes of) phylogenetic trees with n leaves. It is well known [5] that $|\mathcal{T}_1| = |\mathcal{T}_2| = 1$ and $|\mathcal{T}_n| = (2n-5)!! = (2n-5)(2n-7)\cdots 3 \cdot 1$, for every $n \geqslant 3$.

3 Main Result

Let $k, l \in S = \{1, \ldots, n\}$ be any two different labels of trees in $\mathcal{T}_n$. The *distance* $d_T(k, l)$ between the leaves k and l in a phylogenetic tree $T \in \mathcal{T}_n$ is the length of the unique path between them. Let's consider the random variable

$$d_{kl} = \text{distance between the labels } k \text{ and } l \text{ in one tree in } \mathcal{T}_n.$$

The possible values of d_{kl} are $1, 2, \ldots, n-1$.

Our goal is to estimate the value of the median of this variable d_{kl} on $\mathcal{T}_n$ when the tree and the leaves are chosen equiprobably. In this case, $d_{kl} = d_{12}$, and thus we can reduce our problem to compute the median of the variable $d := d_{12}$.

For every $i = 1, \ldots, n-1$, let c_i be the cardinal of $\{T \in \mathcal{T}_n \mid d_T(1,2) = i\}$. Arguing as in [10, p. 140], we have the following result.

Lemma 1. $c_{n-1} = (n-2)!$ *and, for every* $i = 1, \ldots, n-2,$

$$c_i = (n-2)! \frac{(i-1)(n-1)\cdots(2n-i-4)}{(2(n-i-1))!!} = \frac{(i-1)(2n-i-4)!}{(2(n-i-1))!!}.$$

Lemma 2. *For every* $k = 1, \ldots, n-1,$ $\dfrac{1}{(2n-5)!!} \displaystyle\sum_{i=1}^{k} c_i = 1 - \dfrac{2^k(n-3)!(-k+2n-4)!}{2(2n-5)!(-k+n-2)!}.$

Proof. Taking into account that $(2j)!! = 2^j j!$ and $(2j+1)!! = \frac{(2j+1)!}{2^j j!}$, for every $j \in \mathbb{N}$, and using Lemma 1, we have:

$$\frac{1}{(2n-5)!!} \sum_{i=1}^{k} c_i = \frac{(n-3)!}{4(2n-5)!} \sum_{i=2}^{k} \frac{(i-1)2^i(2n-i-4)!}{(n-i-1)!}$$

$$= \frac{(n-3)!}{4(2n-5)!} \sum_{i=1}^{k-1} \frac{i 2^{i+1}(2n-i-5)!}{(n-i-2)!}.$$

We use now the method in [8, Chap. 5] to compute $S_k = \sum_{i=1}^{k-1} \frac{i2^{i+1}(2n-i-5)!}{(n-i-2)!}$.
Set $t_i = i2^{i+1}(2n-i-5)!/(n-i-2)!$. Then

$$\frac{t_{i+1}}{t_i} = \frac{2(1+i)(2+i-n)}{i(5+i-2n)}.$$

The next step is to find three polynomials $a(i), b(i)$ and $c(i)$ such that

$$\frac{t_{i+1}}{t_i} = \frac{a(i)}{b(i)} \cdot \frac{c(i+1)}{c(i)}.$$

We take $a(i) = 2(2+i-n)$, $b(i) = 5+i-2n$ and $c(i) = i$. Next, we have to find a polynomial $x(i)$ such that $a(i)x(i+1) - b(i-1)x(i) = c(i)$. The polynomial $x(i) = 1$ satisfies this equation. Then, by [8, Chap. 5],

$$S_k = \frac{b(k-1)x(k)}{c(k)} t_k + g(n) = \frac{(4+k-2n)2^{k+1}(2n-k-5)!}{(n-k-2)!} + g(n),$$

where g is a function of n. We find this function from the case $k = 2$:

$$\frac{4(2n-6)!}{(n-3)!} = S_2 = \frac{8(6-2n)(2n-7)!}{(n-4)!} + g(n).$$

From this equality we deduce that $g(n) = \dfrac{4(2n-5)!}{(n-3)!}$. We conclude that:

$$S_k = \sum_{i=1}^{k-1} \frac{i2^{i+1}(2n-i-5)!}{(n-i-2)!} = \frac{(4+k-2n)2^{k+1}(2n-k-5)!}{(n-k-2)!} + \frac{4(2n-5)!}{(n-3)!}.$$

The formula in the statement follows from this expression.

Theorem 1. *Let* median(d) *be the median of the variable* d *on* $\mathscr{T}_n$. *Then,*

$$\frac{\text{median}(d)}{\sqrt{4\ln(2)n}} = 1 + O\left(n^{-1/2}\right).$$

In particular, $\displaystyle\lim_{n\to\infty} \frac{\text{median}(d)}{\sqrt{4\ln(2)n}} = 1.$

Proof. To simplify the notations, we shall denote median(d) by $\tilde{k}$. By definition,

$$\tilde{k} = \max\left\{k \in \mathbb{N} \mid \sum_{i=1}^{k} c_i \leqslant \frac{|\mathscr{T}_n|}{2}\right\} = \max\left\{k \in \mathbb{N} \mid \frac{2^k(n-3)!(-k+2n-4)!}{2(2n-5)!(-k+n-2)!} \geqslant \frac{1}{2}\right\}.$$

Thus, $\tilde{k}$ is the largest integer value such that

$$2^{\tilde{k}}(n-3)!(-\tilde{k}+2n-4)! \geqslant (2n-5)!(-\tilde{k}+n-2)!.$$

If we simplify this inequation and take logarithms, this condition becomes

$$\tilde{k}\ln(2) \geqslant \sum_{j=3}^{\tilde{k}+1} \ln\left(\frac{2n-(j+2)}{n-j}\right) = \sum_{j=3}^{\tilde{k}+1} \ln\left(\frac{2-\frac{j+2}{n}}{1-\frac{j}{n}}\right). \tag{1}$$

Combining the development of the function $\ln(\frac{2-(j+2)x}{1-jx})$ in $x = 0$,

$$\ln\left(\frac{2-(j+2)x}{1-jx}\right) = \ln(2) + \frac{1}{2}(j-2)x + \frac{1}{8}(j-2)(3j+2)x^2 + O\left(x^3\right),$$

with equation (1), we obtain:

$$\ln(2) \geqslant \frac{1}{2n}\sum_{j=3}^{\tilde{k}+1}(j-2) + O\left(\frac{\tilde{k}^3}{n^2}\right) = \frac{\tilde{k}(\tilde{k}-1)}{4n} + O\left(\frac{\tilde{k}^3}{n^2}\right).$$

So, the first order term of the median $\tilde{k}$ will be the largest integer value that satisfies $\tilde{k}^2/4n \leqslant \ln(2)$. Therefore, the median will be the closest integer to $\sqrt{4\ln(2)n}$, from where the thesis in the statement follows.

4   Conclusions

We have obtained a limit formula for the median of the distance d between two leaves in a fully resolved unrooted phylogenetic tree with n leaves, namely:

$$\text{median}(d) \sim \sqrt{4\ln(2)n}$$

This value complements the determination of the mean value and the variance of this distance carried out by Steel and Penny [10]. Actually, our method allows to find more terms of the development of the median, with some extra effort. For instance, it can be easily proved that

$$\text{median}(d) \sim \sqrt{4n\ln 2} + (\frac{1}{2} - \ln 2).$$

The limit formula obtained in this work can be generalized to the p-percentile $x_p = \max\left\{k \in \mathbb{N} \mid \Sigma_{i=1}^{k} c_i \leqslant |\mathscr{T}_n| p\right\}$. Indeed, using our method we obtain that $x_p \approx \sqrt{-4\ln(1-p)n}$.

In the near future we plan to extend this work to *arbitrary* unrooted phylogenetic trees, as well as to fully resolved *rooted* phylogenetic trees (for the mean and variance figures in the rooted case, see [6]), and to apply this kind of results in the derivation of a limit formula for the median of the corresponding squared path difference metrics.

Acknowledgements. This work has been partially supported by the Spanish Government, through projects MTM2009-07165 and TIN2008-04487-E/TIN.

References

1. Bluis, J., Shin, D.G.: Nodal distance algorithm: Calculating a phylogenetic tree comparison metric. In: Proc. 3rd IEEE Symp. BioInformatics and BioEngineering, p. 87 (2003)
2. Chang, H., Fuchs, M.: Limit theorems for patterns in phylogenetic trees. Journal of Mathematical Biology 60, 481–512 (2010)
3. Farris, J.: A successive approximations approach to character weighting. Syst. Zool. 18, 374–385 (1969)
4. Farris, J.: On comparing the shapes of taxonomic trees. Syst. Zool. 22, 50–54 (1973)
5. Felsenstein, J.: Inferring Phylogenies. Sinauer Associates Inc., (2004)
6. Mir, A., Rosselló, F.: The mean value of the squared path-difference distance for rooted phylogenetic trees. Submitted to J. Math. Biol (2009),
   http://es.arxiv.org/abs/0906.2470
7. Penny, D., Hendy, M.D.: The use of tree comparison metrics. Syst. Zool. 34(1), 75–82 (1985)
8. Petkovsek, M., Wilf, H., Zeilberger, D.: $A = B$. AK Peters Ltd., Wellesley (1996),
   http://www.math.upenn.edu/~wilf/AeqB.html
9. Phipps, J.B.: Dendrogram topology. Syst. Zool. 20, 306–308 (1971)
10. Steel, M.A., Penny, D.: Distributions of tree comparison metrics—some new results. Syst. Biol. 41, 126–141 (1993)
11. Williams, W.T., Clifford, H.T.: On the comparison of two classifications of the same set of elements. Taxon 20(4), 519–522 (1971)

In Silico AFLP: An Application to Assess What Is Needed to Resolve a Phylogeny

María Jesús García-Pereira, Armando Caballero, and Humberto Quesada

Abstract. We examined the effect of increasing the number of scored AFLP bands to reconstruct an accurate and well-supported AFLP-based phylogeny. *In silico* AFLP was performed using simulated DNA sequences evolving along a symmetric tree with ancient radiation. The comparison of the true tree to the estimated AFLP trees suggests that moderate numbers of AFLP bands are necessary to recover the correct topology with high bootstrap support values (i.e > 70%). However, branch length estimation is rather unreliable and does not improve substantially after a certain number of bands are sampled.

Keywords: AFLP, phylogeny, bootstrap support, simulation, accuracy.

1 Introduction

The amplified fragment length polymorphism (AFLP) technique is becoming extensively used as a source of informative molecular markers for phylogenetic inference in many studies of plants, animals, fungi and bacteria [1]. The technique generates highly reproducible fingerprints that are usually recorded as a 1/0 band presence-absence matrix. Phylogenetic relationships are then inferred analyzing the AFLP matrix directly, or converting it into a distance matrix using dissimilarity measures.

AFLP markers are appropriate for phylogenetic inference as long as sequence divergence is small, the topology of the underlying evolutionary tree is symmetric, and not very short ancestral branches exist [2]. Recent theoretical studies indicate that a major drawback of this technique is the low information content of AFLP markers. This weakness seems to have a much larger negative impact on tree reliability than other commonly invoked limitations of AFLP data sets, such as the occurrence of non homologous bands or the dominant nature of AFLP characters

María Jesús García-Pereira · Armando Caballero · Humberto Quesada
Departamento de Bioquímica, Genética e Inmunología, Facultad de Biología,
Universidad de Vigo, 36310 Vigo, Spain
e-mail: mariajesus@uvigo.es

M.P. Rocha et al. (Eds.): IWPACBB 2010, AISC 74, pp. 137–141, 2010.

[2]. Given that the outcome of any phylogenetic analysis critically depends upon the amount of data available, we focus on determining how many AFLP bands would be needed to resolve a problematic phylogenetic tree. Experimental studies aimed at assessing this question are, however, resource intensive and hampered by the fact that we rarely know the true tree. Here we investigate one potential strategy for improving phylogenetic inference. Using *in silico* AFLP fingerprints, we assess the likelihood that a well supported phylogeny can be resolved using AFLP bands, and how many bands would then be needed.

2 Methods

DNA sequences of 2.8 Mb were generated with the software Seq-Gen [3] using the Jukes and Cantor substitution model. Simulations with Seq-Gen were performed along a phylogenetic tree (hereafter referred as the reference tree) containing 16 sequences, with a symmetric topology and a length from the most internal node to the tip of 0.02 substitutions per site. Branch lengths were specified using an ancient radiation model, in which daughter branches were twice as long as the parent branch. This generated a tree with several short internal branches, making it difficult to resolve.

A computer program written in C was used to simulate the cutting of the generated DNA sequences with restriction enzymes *EcoR*I and *Mse*I, which are the typical enzymes used in AFLP studies. Only fragments sizes between 40 and 440 nucleotides were considered in the subsequent analyses to emulate experimental studies. A combination of selective nucleotides adjacent to the restriction sites was used to simulate the selective amplification of 100 AFLP bands per AFLP profile.

Phylogenies were estimated with PAUP*4 [4] using the minimum evolution method. Estimated AFLP trees were compared with their corresponding reference trees by the program Ktreedist [5]. This program computes a K-score that measures overall differences in the relative branch length and topology of two phylogenetic trees. Topological differences among reference and estimated trees were assessed using Robinson-Foulds (R-F) [6] distance. High K-scores or R-F distances indicate a poor match between the estimated AFLP-based tree and the reference tree. A total of 50 replicates were run per simulation. An average K-score and R-F distance was computed for each set of replicates.

Branch support was determined running 1000 bootstap replicates per simulation. Four different bootstrap consensus trees were estimated with minimum cutoffs of 63%, 86%, 95% and 98% respectively. The average success of resolution [7] was calculated as a weighted average of the four R-F distances resulting from the comparison of the respective consensus trees with the reference tree.

3 Results and Conclusions

All measures of accuracy and support showed a rapid improvement that was subsequently slowed down as an increased number of bands were analyzed

(Figs. 1-2). This is due to the fact that, as clades are resolved, the total number of newly resolved clades decreases for each further increment in band number. High resolution to recover the correct tree topology (R-F distance) was achieved using a total of about 300 bands (Fig. 1). In contrast, branch length estimation (K-score) was rather unreliable across the entire set of band numbers analyzed and did not improve substantially once the amount of about 500 bands was reached.

Our simulations showed that relatively high bootstrap support values (i.e. ≥ 70%) for most nodes were reached when sampling as few as 500 bands, but that an increase in the number of sampled bands did not necessarily equate to a linear increase in support values (Fig. 2). Bootstrap support values ≥ 98% for all nodes required double number of bands, from 500 to 1000.

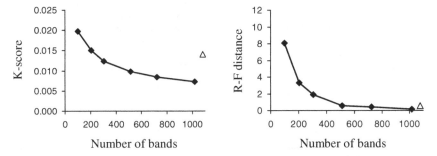

Fig. 1 K-scores and R-F distances resulting from the comparison among the true and estimated minimum evolution trees based on different numbers of AFLP bands. Δ K-score and R-F distance for a 1Kb long DNA sequence

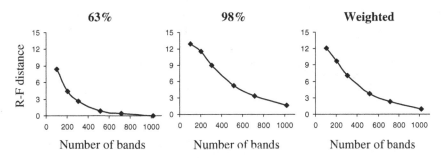

Fig. 2 Bootstrap support. Topological congruence between the reference and estimated consensus trees using different minimum cutoff values and number of sampled AFLP bands

To better visualize the relationship between K-score, R-F distance and tree accuracy, we plotted the reference tree and the estimated AFLP-based trees (Fig. 3). External nodes have a higher bootstrap support than deeper nodes, and nodes achieving 90% bootstrap support were more responsive to increasing the number of sampled bands. Our *in silico* simulations provide new insights into the phylogenetic utility of AFLPs and suggest that moderate amounts of bands (on the range of 300-600) are necessary to recover most clades with high support values.

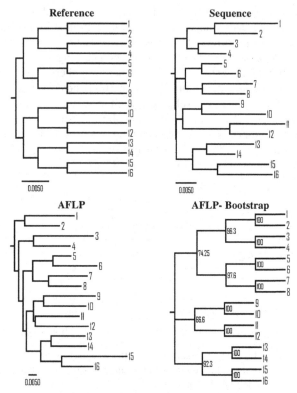

Fig. 3 Reference and estimated trees using DNA sequences 1 Kb long (Sequence) and AFLP characters based on 300 bands (AFLP) and 1000 bands after bootstrap re-sampling (AFLP-Bootstrap). Numbers at the nodes indicate the percentage of bootstrap support

Acknowledgments. This work was funded by the Ministerio de Ciencia e Innovación and Fondos Feder (CGL2009-13278-C02).

References

1. Meudt, H.M., Clark, A.C.: Almost forgotten or latest practice? AFLP applications, analyses and advances. Trends Plant. Sci. 12, 106–117 (2007)
2. García-Pereira, M.J., Caballero, A., Quesada, H.: Evaluating the relationship between evolutionary divergence and phylogenetic accuracy in AFLP data sets. Mol. Biol. Evol. (in press, 2010)
3. Rambaut, A., Grassly, N.C.: Seq-Gen: an application for the Monte Carlo simulation of DNA sequence evolution along phylogenetic trees. Comput. Appl. Biosci. 13, 235–238 (1997)
4. Swofford, D.L.: PAUP*: phylogenetic analysis using parsimony (*and other methods), version 4. Sinauer Associates, Sunderland, Massachusetts (1999)

5. Soria-Carrasco, V., Talavera, G., Igea, J., Castresana, J.: The K tree score: quantification of differences in the relative branch length and topology of phylogenetic trees. Bioinformatics 23, 2954–2956 (2007)
6. Robinson, D.F., Foulds, L.R.: Comparison of phylogenetic trees. Math. Biosci. 53, 131–147 (1981)
7. Simmons, M.P., Webb, C.T.: Quantification of the success of phylogenetic inference in simulations. Cladistics 22, 249–255 (2006)

Employing Compact Intra-genomic Language Models to Predict Genomic Sequences and Characterize Their Entropy

Sérgio Deusdado* and Paulo Carvalho

Abstract. Probabilistic models of languages are fundamental to understand and learn the profile of the subjacent code in order to estimate its entropy, enabling the verification and prediction of "natural" emanations of the language. Language models are devoted to capture salient statistical characteristics of the distribution of sequences of words, which transposed to the genomic language, allow modeling a predictive system of the peculiarities and regularities of genomic code in different inter and intra-genomic conditions. In this paper, we propose the application of compact intra-genomic language models to predict the composition of genomic sequences, aiming to achieve valuable resources for data compression and to contribute to enlarge the similarity analysis perspectives in genomic sequences. The obtained results encourage further investigation and validate the use of language models in biological sequence analysis.

Keywords: language models, DNA entropy estimation, genomic sequences modeling.

1 Introduction

Language models aim to capture the context of a language based on the study and computation of the probabilities of its patterns [1], developing models to infer the rules behind the successions of its segments, i.e. words, n-grams, sounds, codons, etc. Hidden Markov Models (HMMs) also rely on probabilistic models and are widely used in bioinformatics for gene prediction and profiling of sequences [2].

Sérgio Deusdado
CIMO - Mountain Research Centre, Polytechnic Institute of Bragança, Portugal
e-mail: `sergiod@ipb.pt`

Paulo Carvalho
Department of Informatics, School of Engineering, University of Minho, Braga, Portugal

* Corresponding author.

M.P. Rocha et al. (Eds.): IWPACBB 2010, AISC 74, pp. 143–150, 2010.

Entropy measures of DNA sequences estimate their randomness or, inversely, their repeatability [3]. In the field of genomic data compression, fundamentally based on the comprehension of the regularities of genomic language and entropy estimation, language models appear as a promising methodology to characterize the genomic linguistics and to provide predictive models for data compression [4][5] [6], as well as revealing new approaches for sequence similarity analysis [7]. Statistical language models are widely used in speech recognition [8] and have been successfully applied to solve many different information retrieval problems [9]. A good review on statistical language modeling is presented by Rosenfeld in [10].

Currently, the Biological Language Modeling Toolkit is a good example of the interest of this field of investigation, developed by the Center for Biological Language Modeling. This toolkit consists on a compilation of various algorithms that have been adapted to biological sequences from language modeling, and specifically it is oriented to uncover the "protein sequence language". The toolkit is publicly available at the following URL: http://flan.blm.cs.cmu.edu/12/HomePage.

2 Language Models

Language modeling is the art of determining the probability of a word sequence $w_1...w_n$, $P(w_1...w_n)$ [10]. This probability is typically divided into its component probabilities:

$$P(w_1...w_i) = P(w_1) \times P(w_2|w_1) \times ... \times P(w_i|w_1...w_{i-1})$$

(1)

$$= \prod_{i=1}^{n} P(w_i | w_1, w_2, ..., w_{i-1})$$

Since it may be difficult to compute the probability $P(w_i|w_1...w_{i-1})$ for large i, it is typically assumed that the probability of a word depends on only the two previous words. Thus, that trigram assumption can be written as:

$$P(w_i|w_1...w_{i-1}) \approx P(w_i|w_{i-2}w_{i-1})$$

(2)

The trigram probabilities can then be estimated from their counts in a training corpus. Let $C(w_{i-2}w_{i-1}w_i)$ represent the number of occurrences of $w_{i-2}w_{i-1}w_i$ in our training corpus, and similarly for $C(w_{i-2}w_{i-1})$. Then, we can approximate:

$$P(w_i|w_{i-2}w_{i-1}) \approx C(w_{i-2}w_{i-1}w_i) \, C(w_{i-2}w_{i-1})$$

(3)

The most obvious extension to trigram models is to move to higher order n-grams, such as four-grams and five-grams. In genomic language modeling is usual to consider codons as words. Codons are three-letter words from the quaternary genomic alphabet {A, C, G, T}, resulting only 64 possible combinations. Thus, genomic language models generally use higher order n-grams to improve their efficiency.

Smoothing techniques are used to avoid zero probability n-grams which may occur from inadequate training data [11]. In fact, rare trigrams should also integrate the predictive model; therefore its probability, even low, must be greater than zero. On the other hand, smoothing affects high probabilities to be adjusted downward. Not only do smoothing methods generally prevent zero probabilities, but they also attempt to improve the accuracy of the model as a whole.

The most commonly used method for measuring language model performance is *perplexity*. In general, the *perplexity* of a n-gram language model is equal to the geometric average of the inverse probability of the words measured on test data:

$$\sqrt[n]{\prod_{i=1}^{n} \frac{1}{P(w_i \mid w_1...w_{i-1})}} \tag{4}$$

Low *perplexity* of the model means high fidelity predictions. A language model assigning equal probability to 100 words would have perplexity 100. An alternative, but equivalent measure to *perplexity* is entropy, which is simply $\log_2$ of *perplexity*.

3 Developed Work

The developed work corresponds to a framework for entropy estimation and analysis of DNA sequences based on cooperative intra-genomic compact language models. These models will obtain a probability distribution of the next symbol at a given position, based on the symbols previously seen. Based on the experiments of Cao *et al.* [12], we choose to divide our approach into global and local models, combining their contribution to improve the efficiency of the multiparty predictive model. While global models consider the full extension of the analyzed sequences, local models only capture the probabilistic properties of a limited number of bases preceding the base(s) to predict, considering if necessary a variable displacement.

Our aim was to take advantage of the successive probability present mainly in repetitive regions of DNA sequences, as well as in non-repetitive regions where a stochastic model can be efficient too.

We used a backoff n-gram language model [13][14] implemented with arrays of values representing the most probable chain of codons to occur after each one of the 64 possible codons. Our models were not trained based on a corpus because the intention was to apply, subsequently, the resulting framework to an on-line genomic data compression algorithm. In this sense, the resulting compressed file must be self-contained, as the recalculation of probabilities in the decompression process relies only on the data included in the compressed file. Thus, the need for compact models, especially the global model because it is integrated in the compressed file. The local models are adaptive and evolve periodically as they receive new input from the history of the sequence already viewed, i.e. the significant portion of the sequence preceding the point of prediction. In this way, we produce intra-genomic and very compact models, expecting not to compromise the processing time of the predictive algorithm and, additionally, looking forward to include the essential part of the models in the resulting compressed file to help the initial predictions, when history is not available.

Experimental results, using a typical set of DNA sequences (see Table 3) used in DNA compressors as test corpus, showed that ten-grams/codons corresponds to the most appropriated order for global models, considering the tradeoff between model performance and computational cost. For local models, we used order twenty, not codon based but nucleotide based. Each model presents its prediction supported by an associated probability, reflecting the model's confidence. At the end, all local and global models are pooled to elect by a voting system the definitive prediction to be emitted. Independently of the model that produces the prediction, any prediction is just a symbol. Hence, if the prediction is made by a codon-based model only the first nucleotide of the predicted codon is considered.

3.1 Local Models

Local models use single letters (nucleotides) instead of codons and are of order thirty. Being adaptive, they are modified with the latter knowledge obtained from the already predicted - and eventually corrected - sequence. The local models used in our work are based on 1000 nucleotides context, not immediately before the nucleotide to predict but forming a window slid back in the sequence. We used two versions based on different displacements, one with 500 bp displacement and the other going backward 3000 bp. We used different types of local models to enlarge the possibilities of our prediction system, trying to take advantage on the best available knowledge, such as being aware of that most repetitions occur hundreds or thousands of bp after its last occurrence. Considering that some DNA regularities occur in the reverse complementary form, the so-called palindromes, we used complimentary local models based on reverse complementary history.

3.2 Global Models

Global models use codons and gather the probabilistic study of ten-grams. A global model based on the reverse complementary sequence was also produced. Global models are meant to be compact, as they will integrate a future compressed file of the DNA sequence. Our global models are based on tables, containing the most probable succession of codons to occur after each one of the 64 possible codons. Without increasing data complexity, it is possible to calculate global models for the three frames of a sequence. In this way, frame 1, 2 and 3 variants were also considered. These models can be consulted also for subsequences of codons, not necessarily initiated at order 0, using a backing off strategy. An example of a global model, upon analysis of the frame 1 of a sequence, is shown in Table 1.

Table 1 Example of a compact global model considering ten-grams

Frame 1		Order										
		0	1	2	3	4	5	6	7	8	9	10
Code	Codon	Sucession of codons with highest probability										
1	AAA	1→	4	4	16	63	29	32	1	1	2	34
2	AAC	2→	12	12	57	23	44	43	5	4	1	31
...	...	...	...	...	...	...	...	...	...	...	...	...
63	GGT	63→	52	2	64	4	11	13	24	6	12	59
64	GGG	64→	11	13	24	6	12	59	17	55	32	4

3.3 Prediction Process

The test prototype considers six different models, described as follows:

M1 – regular global model;
M2 – global model based on the reverse complementary sequence;
M3 – regular local model considering 1000 previous symbols context and a retro-displacement of 3000 bases;
M4 – reverse complementary local model considering 1000 previous symbols context and a retro-displacement of 3000 bases;
M5 – regular local model considering 1000 previous symbols context and a retro-displacement of 500 bases;
M6 – reverse complementary local model considering 1000 previous symbols context and a retro-displacement of 500 bases.

A model emits a prediction based on order n when it contains the knowledge of a probable n-gram equal to the one at the end of the analyzed portion of the sequence. When there is a conflict between predictions of equal order, global models have priority and their predictions prevail as they derive from the complete sequence. If a global model produces a prediction of order ≥ 3 then the local models predictions are ignored. Each model votes for its predicted symbol, and in the end a probabilistic distribution emerges from a voting system where global models have more weight on final results than local models. Votes from local models are equal to the order used in the prediction, whereas the global models' orders used in the predictions are trebled. Table 2 shows an example of the election of a final prediction and its probability distribution based on the following individual predictions cases:

M1 – predicted A with order 5(x3);
M2 – predicted C with order 1(x3);
M3 – predicted T with order 3;
M4 – predicted A with order 1;
M5 – predicted C with order 2;
M6 – predicted T with order 1;

To prevent zero probability, non-voted symbols receive one vote.

Table 2 Demonstration of the voting system used to achieve the probability distribution

Prediction	Votes by Model						Total	Probability distribution
	M1	M2	M3	M4	M5	M6		
A	15			1			16	62%
C		3			2		5	19%
T			3			1	4	15%
G							1	4%

4 Experimental Results

A test prototype was implemented to combine the predictions from the models described in the previous section in order to assess the predictive capability of our framework. The code was written in C language and compiled using *gcc* version 3.4.2, configured for maximal code optimization. Tests ran on a system based on Intel Pentium IV – 3,4GHz, 8KB L1 + 512 KB L2 cache, with 1GB RAM-DDR and a 250 GB HD. We tested our prototype on a dataset of DNA sequences typically used in DNA compression studies. The dataset includes 11 sequences: two chloroplast genomes (CHMPXX and CHNTXX); five human genes (HUMDYSTROP, HUMGHCSA, HUMHBB, HUMHDABCD and HUMHPRTB); two mitochondria genomes (MPOMTCG and MTPACG); and genomes of two viruses (HEHCMVCG and VACCG).

Table 3 contains the results obtained, considering the percentage of predictions that matched the corresponding symbol in the original sequence.

Table 3 Experimental results

Sequence	Length(bp)	% of correct predictions
CHMPXX	121.024	29
CHNTXX	155.844	30
HEHCMVCG	229.354	27
HUMDYSTROP	38.770	26
HUMGHCSA	66.495	37
HUMHBB	73.323	28
HUMHDAB	58.864	29
HUMHPRTB	56.737	28
MPOMTCG	186.608	27
MTPACG	100.314	27
VACCG	191.737	28
Average	**116.279**	**29**

Considering the quaternary alphabet and carrying out a random prediction, it will be expectable, in theory, to obtain 25% of correct predictions, on average. Comparatively, the obtained results exhibit 29% of prediction correctness on average. However, the obtained results are satisfactory considering only intra-genomic study and the reduced size of the used models; moreover the high level of entropy inherent to DNA sequences justifies the quality of the results. HUMGHCSA is the sequence where our predictive model performed better because it is, within the tested set of sequences, the one with lower entropy, i.e. high redundancy, as we may confirm in the literature [4][15].

5 Conclusions and Future Work

Experimental results demonstrate a linear correlation facing the entropy of each tested sequenced based on the results of existing DNA data compressors [4][3][12][15]. Sequences with higher levels of entropy are more difficult to model and hence our models get modest results on their analysis. Global models capture the most significant patterns of the sequence and perform generally better. Local models revealed low utility in entropy estimation but they are important to complement predictions. Different sequences may need proper adjustments of the extension and the displacement of the local models to optimize their prediction capability. We believe that it would be useful to determine the profile of the sequence in advance in order to adaptively adjust the local model's parameterization. This will be addressed in future developments.

Our major goal was to test the potential and efficiency of language models as a complementary compression method for biological data compression, for instance, to complement dictionary based techniques. Consequently, our focus was mainly in regions of the sequences where linear patterns are sparse or exist repeatedly but with reduced extension. Additional work is needed to optimize all the models, especially the local ones, but the obtained results encourage further investigation.

References

1. Durbin, R., Eddy, S., Krogh, A., Mitchison, G.: Biological sequence analysis: probabilistic models of proteins and nucleic acids. Cambridge University Press, Cambrige (1998)
2. Koski, T.: Hidden Markov Models for Bioinformatics. Kluwer Academic Publishers, Dordrecht (2001)
3. Lanctot, J.K., Li, M., Yang, E.: Estimating DNA sequence entropy. In: SODA 2000: Proceedings of the eleventh annual ACM-SIAM symposium on Discrete algorithms, pp. 409–418. Society for Industrial and Applied Mathematics, San Francisco (2000)
4. Loewenstern, D., Yianilos, P.N.: Significantly Lower Entropy Estimates for Natural DNA Sequences. In: Data Compression Conference (DCC 1997), p. 151 (1997)
5. Osborne, M.: Predicting DNA Sequences using a Backoff Language Model (2000), http://www.cogsci.ed.ac.uk/~osborne/dna-backoff.ps.gz
6. Venugopal, K.R., Srinivasa, K.G., Patnaik, L.M.: Probabilistic Approach for DNA Compression. In: Soft Computing for Data Mining Applications, pp. 279–289. Springer, Heidelberg (2009)
7. Buehler, E.C., Ungar, L.H.: Maximum Entropy Methods for Biological Sequence Modeling. In: Workshop on Data Mining in Bioinformatics (with SIGKDD 2001 Conference), San Francisco, CA, USA, pp. 60–64 (2001)
8. Jelinek, F.: Statistical methods for speech recognition. MIT Press, Cambridge (1997)
9. Zhai, C.: Statistical Language Models for Information Retrieval. Synthesis Lectures on Human Language Technologies 1, 1–141 (2008)
10. Rosenfeld, R.: Two Decades of Statistical Language Modeling: Where Do We Go from Here? Proceedings of the IEEE 88, 1270–1278 (2000)

11. Chen, S.F., Goodman, J.: An Empirical Study of Smoothing Techniques for Language Modeling. Harvard University (1998)
12. Cao, M.D., Dix, T.I., Allison, L., Mears, C.: A Simple Statistical Algorithm for Biological Sequence Compression. In: 2007 Data Compression Conference (DCC 2007), Snowbird, UT, USA, pp. 43–52 (2007)
13. Katz, S.: Estimation of probabilities from sparse data for the language model component of a speech recognizer. IEEE Transactions on Acoustics, Speech, and Signal Processing 35, 400–440 (1987)
14. Kneser, R., Ney, H.: Improved backing-off for m-gram language modeling. In: IEEE Int. Conf. Acoustics, Speech and Signal Processing, pp. 181–184. IEEE, Detroit (1995)
15. Korodi, G., Tabus, I.: An efficient normalized maximum likelihood algorithm for DNA sequence compression. ACM Transactions on Information Systems 23, 3–34 (2005)

Structure Based Design of Potential Inhibitors of Steroid Sulfatase

Elisangela V. Costa, M. Emília Sousa, J. Rocha, Carlos A. Montanari, and M. Madalena Pinto[*]

Abstract. The enzyme steroid sulfatase (STS) activity is high in breast tumors and elevated levels of STS mRNA expression have been associated with a poor prognosis. Potent STS irreversible inhibitors have been developed, paving the way to use this new type of therapy for breast cancer. Synthetic small molecules belonging to a focused library of inhibitors of tumor cell growth already obtained and new molecules planned to be reversible inhibitors of STS were docked into STS using the program AutoDock 4. To guide the docking process of the select ligands through the lattice volume that divides the receptor's area of interest, a full set of grid maps was built using the program AutoGrid. Some of the new designed small molecules showed calculated binding affinity for STS presenting ΔG values in a range of -11.15 to -13.07 kcal.mol^{-1}. The synthesis of the most promising STS inhibitors, based on these results, is in progress.

1 Introduction

The highest frequency of breast cancer is observed in postmenopausal women, associated with high levels of peripherical estrogens produced in situ. The enzyme steroid sulfatase (STS) is responsible for conversion of inactive sulfate-conjugated steroids to active non-conjugated forms (Reed et al. 2005). This phenomenon plays a crucial role in the development of the breast cancer hormone-receptor-positive.

Elisangela V. Costa · M. Emília Sousa · M. Madalena Pinto
Department of Chemistry, Laboratory of Organic and Pharmaceutical Chemistry, Faculty of Pharmacy, University of Porto, Rua Aníbal Cunha, 164, 4050-047 Porto, Portugal
e-mail: madalena@ff.up.pt

Elisangela V. Costa · M. Emília Sousa · M. Madalena Pinto
CEQUIMED-UP, Center of Medicinal Chemistry – University of Porto, Rua Aníbal Cunha, 164, 4050-047 Porto, Portugal

J. Rocha · Carlos A. Montanari
Grupo de Estudos em Química Medicinal de Produtos Naturais – NEQUIMED-PN, Instituto de Química de São Carlos, Universidade de São Paulo, São Carlos, SP, Brasil

[*] Corresponding author.

M.P. Rocha et al. (Eds.): IWPACBB 2010, AISC 74, pp. 151–156, 2010.
springerlink.com   © Springer-Verlag Berlin Heidelberg 2010

STS inhibitors can be classified as reversible and irreversible. Most of the reversible inhibitors are substrate- or product-based, feature a steroid skeleton, and are generally less potent than the irreversible blockers (Nussbaumer and Billich 2004). Attempts to design nonsteroidal STS inhibitors have revealed the benzophenone-4,4'-*O,O*-bis-sulfamate (BENZOMATE) as a potent STS irreversible inhibitor (Hejaz et al. 2004). Structure–activity relationship studies showed that the carbonyl group is pivotal for activity and also that the *bis*-sulfamate moiety is responsible for the irreversible mechanism of action.

Until very recently the field has been dominated by irreversible, arylsulfamate-based inhibitors, although a stable, potent reversible inhibitor should be less problematic for development. The main reason apparently is that the inhibitor design was hampered by the lack of the 3D structure of STS and that the discovery of novel inhibitor types has been limited to high-throughput screening (HTS). Nonetheless, the structure of STS has already been determined at 2.60 Å resolution by X-ray crystallography (Hernandez-Guzman et al. 2003).

These facts led us to investigate potential STS reversible inhibitors by structure-based design (SBS), with a benzophenone and a xanthone scaffold (Figure 1). The SBS is a molecular docking process that takes small molecule structures from a database of existing compounds (or of compounds that could be synthesized), and docks them into the protein-binding site, which involves the prediction of ligand conformation and orientation (or posing) within the target (Reich et al. 1992).

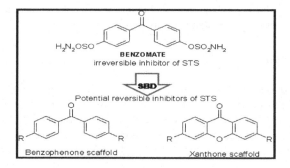

Fig. 1 Strategy used in this work

In order to proceed with these kinds of studies, the strategy used was the docking of compounds with a benzophenone and a xanthone scaffold provided of a significantly similarity in their structure to BENZOMATE (Figure 1). So, small molecules obtained in our group which have been already described as tumor cell growth inhibitors (Pedro et al. 2002, Sousa et al. 2002, Castanheiro et al. 2007, Sousa et al. 2009), as well as new benzophenone/xanthone derivatives designed to be reversible inhibitors of STS, i.e., without the sulfamate group, were docked into STS. The synthesis of the designed xanthones which showed high affinity for STS was already initiated.

2 Docking Studies

An automated docking method, AutoDock 4 (Morris et al. 1998, Morris et al. 2009), was used that predicts the bound conformations of flexible ligands to macromolecular targets in combination with a new scoring function that estimates the free energy change upon binding. AutoDock 4 program allows fully flexible modeling of specific portions of the protein, in a similar manner as the flexible ligand (Morris et al. 2009) only applied for flexible designed molecules that therefore have potentially a reversible mode of action and not for docking of covalently-attached complexes, irreversible ones. AutoGrid calculated the non-covalent energy of interaction between the receptor and a probe atom located in the different grid points of a lattice that divides the receptor's area of interest, the active site, in a grid box. The region to be explored was chosen and delimited by a three-dimensional box so that all important residues forming the active site could be taken into consideration during the docking process (red-colored region on Figure 2).

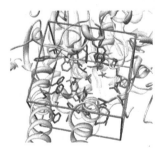

Fig. 2 Grid box

Docking using AutoDock 4 allowed the prediction of the bound conformations of flexible ligands to the macromolecular target. For instance, a xanthone was docked into STS with a good value of ΔG=-12.50 kcal.mol-1 and showed to be well accommodated by a variety of hydrophobic residues such Phe, Val and Leu, which are predominant in STS active site. Moreover, the predicted structure for the ligand-receptor complex suggests two hydrogen bond interactions involving Arg98 and carbonyl group of xanthone scaffold (Figure 3). The cluster analysis shown in Figure 4 of the docked xanthone demonstrates that probably the conformation adopted is the one with a value of $\Delta G = -12.50$ kcal.mol^{-1} (red colored).

The 61 compounds belonging to a focused library of inhibitors of tumor cell growth already obtained were docked into the STS. The respective values of ΔG were determinate in a range of –5.39 to –9.40 kcal.mol^{-1}. Two potent inhibitors against MCF-7 breast adenocarcinoma with the lowest GI_{50} values of 6.0 and 21.9 μM were associated with favorable ΔG values (-8.58 and -6.40 kcal.mol^{-1}, respectively). Some of the new designed xanthones/benzophenones showed high affinity for STS with ΔG values in a range of -11,15 to -13.07 kcal.mol^{-1}. It is possible to observe in Figure 5 the high shape complementarities of two designed molecules

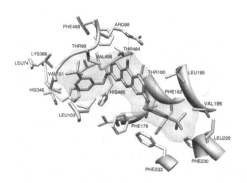

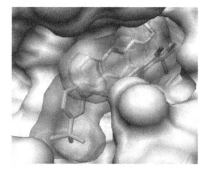

Fig. 3 Interactions suggested by the predicted structure for the ligand-receptor complex

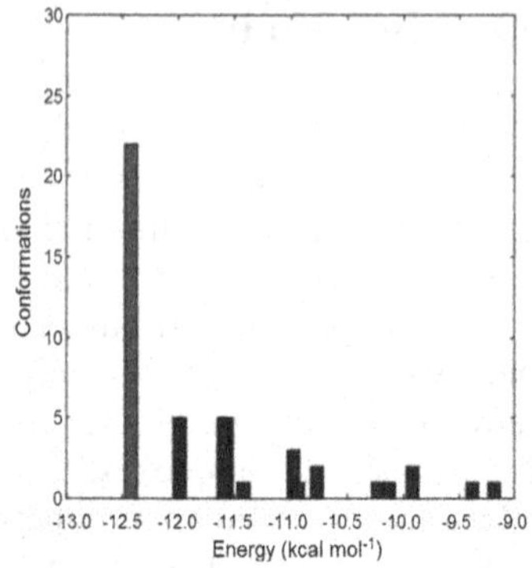

Fig. 4 Cluster analysis of the docked xanthone

Fig. 5 Superimposition of the best conformations of two molecules docked

with the buried active site of STS. The molecules can adopt conformations that are able to fill the whole cavity and prevent the substrate to access to the catalytic residues.

3 Synthesis

The synthesis of the designed molecules associated to the best results was initiated according to conditions in Figure 6. The coupling reagent, TBTU (*O*-(benzotriazol-1-yl)-*N,N,N',N'*-tetramethyluronium tetrafluoroborate), often used in the peptide synthesis, was applied in the esterification of three carboxylic acids with 3,6-dihydroxyxanthone to afford the corresponding esters (Figure 6A). The introduction of bulky groups was also performed in order to obtain more two derivatives (Figure 6B).

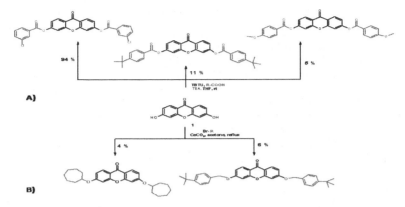

Fig. 6 Ongoing synthesis of new xanthone derivatives potential inhibitors of STS

4 Conclusions and Future Work

Some molecules synthesized in our group already described as tumor cell growth inhibitors were associated to favorable ΔG values. It is expected that those ones might inhibit the tumor cell growth presumably by inhibiting STS. The STS activity of these compounds and the new compounds with favorable ΔG values will be assessed experimentally in the STS assay. We expect that the virtual screening allied to the structure-based design provides new STS reversible inhibitors belonging to xanthone/benzophenone families.

Acknowledgments

Fundação para a Ciência e a Tecnologia (FCT), Unidade I&D 4040/2007, FEDER, POCI for financial support and for the PhD grant to Elisangela Costa (SFRH/BD/30615/2006).

References

Castanheiro, R., Pinto, M., Silva, A., et al.: Dihydroxyxanthones prenylated derivatives: synthesis, structure elucidation, and growth inhibitory activity on human tumor cell lines with improvement of selectivity for MCF-7. Bioorg. Med. Chem. 15, 6080–6088 (2007)

Hejaz, H., Woo, L., Purohit, A., et al.: Synthesis, *in vitro* and *in vivo* activity of benzophenone-based inhibitors of steroid sulfatase. Bioorg. Med. Chem. 12, 2759–2772 (2004)

Hernandez-G, F., Higashiyama, T., Pangborn, W., et al.: Structure of Human Estrone Sulfatase Suggests Functional Roles of Membrane Association. J. Biol. Chem. 278, 22989–22997 (2003)

Morris, G., Goodsell, D., Halliday, R., et al.: Automated Docking Using a Lamarckian Genetic Algorithm and an Empirical Binding Free Energy Function. J. Comput. Chem. 19, 1639–1662 (1998)

Morris, G., Huey, R., Lindstrom, W., et al.: AutoDock 4 and AutoDock tools 4: automated docking with selective receptor flexibility. J. Comput. Chem. 00, 1–7 (2009)

Nussbaumer, P., Billich, A.: Steroid Sulfatase Inhibitors. Med. Res. Rev. 24, 529–576 (2004)

Pedro, M., Cerqueira, F., Sousa, E., et al.: Xanthones as Inhibitors of Growth of Human Cancer Cell Lines and Their Effects on the Proliferation of Human Lymphocytes *In Vitro*. Bioorg. Med. Chem. 10, 3725–3730 (2002)

Reed, M., Purohit, A., Woo, L., et al.: Steroid sulfatase: molecular biology, regulation, and inhibition. Endocrine Rev. 26, 171–202 (2005)

Reich, S., Fuhry, M., Nguyen, D., et al.: Design and synthesis of novel 6,7-imidazotetrahydroquinoline inhibitors of thymidylate synthase using interactive protein crystal structure analysis. J. Med. Chem. 35, 847–858 (1992)

Sousa, E., Paiva, A., Nazareth, N., et al.: Bromoalkoxyxanthones as promising antitumor agents: Synthesis, crystal structure and effect on human tumor cell lines. Eur. J. Med. Chem. 44, 3830–3835 (2009)

Sousa, E., Silva, A., Pinto, M., et al.: Isomeric Kielcorins and Dihydroxyxanthones: Synthesis, Structure Elucidation, and Inhibitory Activities of Growth of Human Cancer Cell Lines and on the Proliferation of Human Lymphocytes In Vitro. Helv. Chim. Acta. 85, 2862–2876 (2002)

Agent-Based Model of the Endocrine Pancreas and Interaction with Innate Immune System

Ignacio V. Martínez Espinosa, Enrique J. Gómez Aguilera,
María E. Hernando Pérez, Ricardo Villares, and José Mario Mellado García

Abstract. In the present work we have developed an agent-based model of the interaction between the beta cells of the endocrine pancreas and the macrophages of the innate immune system in a mouse. The aim is to simulate the processes of proliferation and apoptosis of the beta cells and the phagocytosis of cell debris by macrophages. We have used data from the literature to make the model architecture and to define and set up the input variables. This model obtains good approximations to the real processes modeled and could be used to shed light on some open questions about phagocytosis, wave of apoptosis in the young mice, growing of the beta cell mass and processes that could induce the immune response against beta cells related to type 1 diabetes.

1 Introduction

Diabetes mellitus is a metabolic imbalance characterized by high levels of glucose caused by a deficiency in the secretion or the action of insulin. Type 1 diabetes is the result of the destruction of the cells in the endocrine pancreas that secrete insulin (the beta cells) inside Langerhans islets caused by an immune attack (autoimmune response) against them. The treatment consists of infusion of exogenous

Ignacio V. Martínez Espinosa · Enrique J. Gómez Aguilera · María E. Hernando Pérez
Grupo de Bioingeniería y Telemedicina, Departamento de Tecnología Fotónica,
Universidad Politécnica de Madrid, Madrid, España
e-mail: {ivme, egomez, elena}@gbt.tfo.upm.es

Ignacio V. Martínez Espinosa · Enrique J. Gómez Aguilera · María E. Hernando Pérez
Centro de Investigación Biomédica en Red en Bioingeniería,
Biomateriales y Nanomedicina CIBER-BBN, Madrid, España

Ricardo Villares · José Mario Mellado García
Departamento de Inmunología y Oncología. Centro Nacional de Biotecnología/CSIC,
Madrid, España
e-mail: {rvillares, mmellado}@cnb.csic.es

M.P. Rocha et al. (Eds.): IWPACBB 2010, AISC 74, pp. 157–164, 2010.
springerlink.com   © Springer-Verlag Berlin Heidelberg 2010

insulin in order to control glucose levels depending on food intake [1]. Some efforts nowadays are focusing on creating new therapies to solve the immunological problem and the present work is an approach on applying computational models to this. A model of a real system can be used to comprehend the entire or part of the processes, to predict actions of the system depending on stimuli and to control those stimuli or inputs knowing the system responses [2]. The *in-silico* experiments would be an example about the feature of prediction. Agent-based models like the one that we propose in this paper are a good approach to cell systems because they are composed of discrete elements with particular actions that interact with each other generating a complex behavior [3]. Agent-based models have been used for modeling different problems related to cellular systems and the immune system like the one we propose [4, 5, 6, 7, 8].

The autoimmune response against beta cells needs the loss of control in different steps. The first step is the activation of the innate immune system by the beta cells [9]. This stimulus can be provoked by the necrotic death of beta cells or the apoptotic death followed by a secondary necrosis [10]. The second step is the antigen presentation to T helper lymphocytes (CD4+) so there must be autoreactive lymphocytes to beta cell self-antigens [9]. And the third step consists of a failure of the peripheral tolerance mechanisms mainly due to a lack of regulatory CD4+ lymphocytes action [11]. The present work focuses on the first step, with the proliferation, apoptosis and growing number of the beta cells.

Beta cells in the pancreas carry out proliferation when there's need to maintain or increase the number of cells and apoptosis when some cells need to be substituted. There's a process called *wave of apoptosis* which implies a big loss of beta cells by apoptosis around the time of weaning in mice [12]. This process of beta cell apoptosis has been proposed as the first moment of autoimmune stimulation [12, 13]. A feature detected in innate immune system that is differential in diabetic mice is reduced macrophage phagocytosis rates [13]. There are mathematical approaches to model these first interactions among beta cells and the innate immune system [13]. The present work is a contribution to integrate these characteristics of the beta cells in the pancreas and the first processes of the type 1 diabetes using an agent-based approach with the abstraction of 1 cell – 1 agent.

2   The Model

2.1   Overview

The model simulates a fragment of the pancreas, defined by an initial volume, V_c. This simulated compartment is composed of a number of islets of Langerhans that will be containers of beta cell agents and macrophage agents and manage the influx and efflux of agents. Time is represented as discrete entity. The agents will have different states. Beta cell agent can be on 4 different states: resting, proliferating or carrying out the mitosis process, apoptotic and necrotic. The total number of the beta cell agents in each state in the whole model will be denoted as B_r, B_p, B_a and B_n respectively. $B_f = B_r + B_p$ will represent the total amount of functional beta cells in the compartment. Macrophage agent's states are 2: resting and activated, and the total number of them will be denoted as M_r and M_a. Those numbers of agents will vary during the simulation time and will be taken into account as results of the model. All agents can access population numbers and macrophage agent can set a beta cell agent which is inside the same islet to be eliminated from the simulation, emulating phagocytosis. The simulation time emulates the natural time of the mouse life since birth. All the agents make their actions in each time step and the interval between steps will be denoted as ΔT. Agent actions are deciding to proliferate, phagocyte, life time discount, etc and will be explained further. No agent-based modeling platform was used, the model was implemented directly in java and the graphical results displayed using MatLab. Next, we'll explain the different components of the model and the associated input parameters.

2.2 Compartment and Langerhans Islets

The initial sizes (in cell number) of the Langerhans islets in the compartment will be defined by the Weibull probability distribution that approximates the real size distribution [14]. The density function of this distribution is:

$$P(s) = \frac{\gamma}{\eta}\left(\frac{s}{\eta}\right)^{\gamma-1}\exp\left[-\left(\frac{s}{\eta}\right)^{\gamma}\right]$$

Where γ is called the shape parameter and η is the scale parameter. The result of the distribution is s, a measurement of islet volume by the beta cell number, n: $n=s^3$. γ doesn't vary much during time and we take the estimated in [14] as 1.64. η is based on the value for adult mice: 10.15 according to [14]. If the mean islet size is composed of 1800 beta cells in adult and the beta cell number is proportional to body weight after wave of apoptosis [15], the total beta cell loss during this process is 60% and the ratio of average weight of newborn and adult mice is 1.25g/25g=0.05 [16], we estimate the mean islet size for newborn mice as 0.05·1800/(1-0.60)=225 beta cells. The mean size using the Weibull function is $n=\eta^3\cdot\Gamma(1+3/\gamma)$, using $n=225$ and $\gamma=1.64$, the value of η is approximately 5.

We take the volume of a beta cell (V_b) as inverse of beta cell density in an islet, $V_b = 1/(4\cdot10^8 \text{ cells/ml}) = 2.5\cdot10^{-9}$ ml [13]. Initial beta cell density (D_i) in the pancreas is estimated by de 1% of the volume in adults [15]: $0.01\cdot4\cdot10^8$ cells/ml=$4\cdot10^6$ cells/ml, taking that the density in newborn is 3.5 times larger than in adults for humans [17] and assuming for mice similar values, the initial density is $3.5\cdot4\cdot10^6$ cells/ml = $14\cdot10^6$ cells/ml of pancreatic volume.

The wave of apoptosis will be defined as a function of time of the proportion of beta-cell dead in each time step. We assume that the 60% of the beta cells die during the wave with a peak rate of 9% per day around the 15th day of life [12]. We approximate a Gaussian outline for the function of the wave and we have this:

$$W_p = \exp\left[-\left(\frac{t-15}{0.6/(0.09\sqrt{\pi})}\right)^2\right] \quad ; \quad W_I = \int_0^t W_p\,dt$$

We use average mouse body weight from newborn to adult M_{bw} [16] (fig. 1.B). The pancreas mass, P_m, has been estimated as 1% of body weight [18].

Number of beta cells to maintain in the compartment (B_{man}): This is the number that the beta cells agents of the model will need to reach at each time step by undergoing cell division. It's calculated as $B_{man} = B_{t0} \cdot (M_{bw}/M_{bw0}) \cdot (1 - W_l)$, where B_{t0} is the initial number of functional beta cells in the compartment and M_{bw0} is the initial average mice body weight (1.25g).

2.3 Beta Cell Agent

The characteristics of the beta cell agent are the life span since it's created as a result of mitosis until it suffers apoptosis, mitosis duration, proliferation dynamics and duration of apoptotic and necrotic debris.

We estimate for mice in 50 days of mean life span [15] and with an exponential probability distribution. Mitosis duration is about 6 hours [15] and cell cycle time for a beta cell is 14.9 hours [15] so there's 14.9–6=8.9 hours of refractory time.

Proliferation dynamics is based on the number of B_{man} to maintain at each time step. All the new beta cells produced will come from the mitosis of the previous [19, 20] and the probability of entering in mitosis of a beta cell is independent of the others in the same or different islets [21]. This dynamics is modeled as a probability of entering in mitosis, p_m, in each time step. This probability is separated in two addends: A basal probability that would maintain the beta cell number without gains or losses: p_b. And a growing probability, p_g, if B_{man} exceeds B_t. $p_m = p_b + p_g$, where p_b is the result of inverse of life span scaled by ΔT: $p_b = (1/BL) \cdot \Delta T$; and p_w is calculated by the unitary difference between B_{man} and B_t scaled by ΔT too: $p_w = [(B_{man} - B_t)/ B_t] \cdot \Delta T$. There's a maximum for this probability for limiting the % of beta cells in mitosis. In [22] they kill 75% of the beta cells of a mouse at 5 weeks of age. They calculate de % of beta cells in mitosis in these mice (8.5%) and controls (2%). The probability of proliferation is 8.5/2 = 4.25 times bigger. If the rate of beta cell proliferation per day at 5 weeks of age is 5-7% [15] (we take 6%, 6%/2%=3 times bigger than the probability of finding a mitotic cell with the procedures of [22]), then the maximum proliferation rate per day will be 3·8.5% = 25.5% and approximated to 25.5%·ΔT per time step. Finally, duration of apoptotic debris before suffering secondary necrosis is 2 days [10] and duration of necrotic debris is estimated in our model to 10 days.

2.4 Macrophage Agent

Firstly, there's a macrophage influx rate for reaching the islets. The rate used in [13] is M_i=50000 macrophages/(ml·day) but this rate in our model is underestimate if we calculate the influx for each islet applying the islet volume (V_{is}) as number of beta cells multiplied by V_b. Instead of this we assume a volume wider that double of the islet radius (10 times V_{is}) and we take absolute macrophage influx as $10 \cdot M_i \cdot V_{is}$. We use the estimated mean life span inside the volume (M_v in our case) by [13] as a mean of 10 days with exponential probability distribution.

The phagocytosis dynamic is modeled as a rate (P_r) of dead beta cells engulfed per unit of time. This rate is not constant because it is know that phagocytosis is enhanced after previous debris intakes [13] so we model this rate as a function of an *activation level* (A_l) and its value will be among a minimum (P_{r_min}) and a maximum (P_{r_max}). $P_r = P_{r\_min} + p_{act} \cdot A_l$, where p_{act} is the parameter that relates linearly A_l with P_r. In [23] they find that the maximum P_r is reached after first engulfment. A_l increases one unit per engulfed, then $p_{act} = P_{r\_max} - P_{r\_min}$.

The deactivation rate has an exponential decay of 0.4 per day ($M_a = M_{a0} \cdot e^{-0.4t}$) [23] so we assume a loss of A_l with the same exponential decay and we consider activated macrophage when its A_l is over e^{-1} and resting when it is lower. In [23] and [13] they determine a basal phagocytosis rate of $2 \cdot 10^{-5}$ ml/(cell·day) and a maximal phagocytosis rate of 5.10^{-5} ml/(cell·day) for BALB/c mice (healthy mice) and a basal and maximal phagocytosis rate of $1 \cdot 10^{-5}$ ml/(cell·day) for NOD mice (diabetic). Scaling to $4 \cdot 10^{8}$ beta cells/ml [13] we have for a BALB/c macrophage: P_{r_min}=8000 cells/day, P_{r_max}=20000 cells/day and p_{act}=12000 cells/(day·A_l) and for a NOD macrophage: P_{r_min}=4000 cells/day, P_{r_max}=4000 cells/day and p_{act}=0 cells/(day·A_l). The density $4 \cdot 10^{8}$ cells/ml includes all the beta cells so the macrophages contact with P_r cells/day and engulf only the apoptotic or necrotic cells, marking them for removing from the islet. Maximum number of engulfed cells is 7 per macrophage and the digestion rate is 25 cells/day [23].

3 Results and Discussion

Apart from the number of agents in different states and B_{man}, we calculate other rates: percent of beta cells in mitosis (B_{mp}=100·B_p/B_t), percent of beta cells that proliferates per day (B_{pr}=$100 \cdot (B_{pn}/B_t) \cdot \Delta T$, where B_{pn} is the number of new beta cell in mitosis in each time step), number of necrotic cells phagocyted (NB_{nf}), total number of beta cells in the organism (NB_t) and beta cell number per body weight ($NB_m = NB_t / M_c$). It's proved that there is no islet neogenesis since early in mouse life [19, 20]. For obtaining NB_t, in our model, we assume that there are no new islets created after birth, then, the whole amount of islets in the pancreas is proportional (K) to the initial volume of the compartment (V_c) and the initial volume of the pancreas (V_{p0}): $K = V_{p0}/V_c$, and the same proportion applies for beta cell number. We assume that the density of the pancreas is nearly the water density (d_p=1g/ml), then V_{p0}=P_{m0}/d_p (P_{m0} is the initial pancreatic mass), and finally, the initial NB_t=$K \cdot B_{t0}$. We have run simulations with this model for a life length of 50 days and ΔT=1 hour. V_c=10^{-4} ml so B_{t0}=$D_i \cdot V_c$=1400 cells. The results are presented in figure 1. The parameters used for phagocytosis are taken for NOD except in figure 1.E that are taken for NOD and BALB/c.

B_{man} (fig. 1.A) present a profile with a phase of growth after birth, a plateau around the days of the wave of apoptosis and finally another phase of growing. This profile of B_{man} is proportional to the profile of NB_t (fig. 1.B) and simulates the real growth dynamics of beta cell mass in the real pancreas of mice [24] where we can see the first growth, the plateau between 10 and 20 days of life with even a

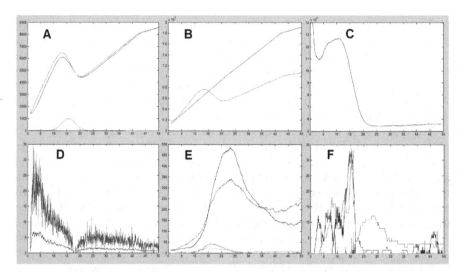

Fig. 1 Results. A) B_{man} (green), B_t (blue), $W_p \cdot 1000$/peak (red). B) M_c[grams]$\cdot 10^5$ (blue), NB_t (green). C) NB_m. D) B_{pr} (green), B_{mp} (blue). E) M_r for NOD (blue), M_r for BALB/c (red), M_a for NOD (green) and M_a for BALB/c (cyan). F) B_a (blue), B_n (green), NB_{nf} (red)

slight decrease, and finally the final growth. This final growth is proportional to body weight in the model and it is the same result who researchers find in muridae after the first month of life [15, 25]. The proliferative dynamics of beta cells follows the conditions of independency [21] and non neogenesis [19, 20] in the model and B_t value follows B_{man} value (fig. 1.A). The quantification of the rate B_{pr} in real muridae finds values of 18% at birth, 7% at one month of age and 2-3% during adulthood [15]. In our model (fig. 1.D) we can see this rate, for the same moments as 22%, 5% and 2-3%, very similar values that reproduce the real proliferation dynamics. This decreasing proliferation rate with age is also seen in [12] but with lower values at birth, and if we take into account some result that report percents of beta cells in mitosis using different techniques (BrdU [25] and Ki67 antibody [22] staining protocols) for 1 month of age the results are consistent with the model too. The values of NB_m after apoptosis wave are high estimated compared with real studies: about 50000 beta cells per gram in the model (fig. 1.C) and around 30000 cells/g in [18, 22], those studies find 0.06-0.07 mg of beta cell mass per gram of body weight and if we assume density of 1g–1ml: $0.07 \cdot 10^{-3}/V_b$ is about 30000 cells/g. This difference could come from overestimate D_i. The loss of beta cell mass when it's the onset of type 1 diabetes is 40% for adults and 85% near birth in humans [17]. A minimum threshold of NB_m has to exist to maintain the levels of insulin per gram of body weight. Extrapolating these findings to our mouse model, the (100%-40%) of our beta cell mass per gram for adults is about $50000 \cdot 0.6 = 30000$, it would be the NB_m threshold for diabetes. At birth, NB_m is around 130000 cells/g. To reach the threshold (30000 cells/g), the loss of beta cell mass would need to be of 77%, a value quite similar to the 85% found in [17].

The macrophage profile found for the different phagocytosis rates for BALB/c and NOD mice shows a high influx for the last one with more number in activated state (fig 1.E), consistent with [13] results and supporting [9] theory about higher and prolonged activation rates of innate immune system that could even lead to a positive feedback of beta cell destruction and immune activation that would lead to type 1 diabetes. The amount of B_a present at early ages is higher than in older mice, another model result (fig. 1.F) consistent with experimentation [15, 25], and a higher NB_{nf} during first weeks, highlighting this period as the more prone to immune activation.

4 Conclusions

The model presented achieves the reproduction of beta cell mass increase dynamics, proliferation rates, beta cells in mitosis percent and descent of the number of beta cells per body weight. Therefore, this model represents a connection among scientific data in the literature. Hence, it can be considered when it comes to make estimates where variations of parameters represented by inputs in the model could be motif of study or experimentation.

Furthermore, this model obtains different responses of macrophage influx for NOD (diabetic) and BALB/c (healthy) mice depending on phagocytosis rates during wave of apoptosis. It also finds more contacts with necrotic debris during early life. These results together support the simulation as a predictor of the initials events for the development of autoimmune response of type 1 diabetes and present the agent-based modeling applied to this pathology as a good tool for comprehension and prediction of the processes that lead to its onset.

References

1. Porth, C.M.: Pathophysiology: Concepts of altered health states, 7th edn. Lippincott Williams & Wilkins, USA (2007)
2. Haefner, J.W.: Modeling Biological Systems: Principles and Applications, 2nd edn. Springer, New York (2005)
3. Macal, C.M., North, M.J.: Tutorial on Agent-Based Modeling and Simulation part 2: How to Model with Agents. In: IEEE Proceedings of the 2006 WSC, pp. 73–83 (2006)
4. Duca, K.A., Shapiro, M,, Delgado-Eckert, E., et al.: A Virtual Look at Epstein–Barr Virus Infection: Biological Interpretations. PLoS Pathog 3(10), e1388–e1400 (2007)
5. Folcik, V.A., An, G.C., Orosz, C.G.: The Basic Immune Simulator: An agent-based model to study the interactions between innate and adaptive immunity. Theor. Biol. Med. Model 4, 39 (2007)
6. Ballet, P., Tisseau, J., Harrouet, F.: A multiagent system to model an human humoral response. IEEE, Los Alamitos (1977), doi:10.1109/ICSMC.1997.625776
7. Baldazzi, V., Castiglione, F., Bernaschi, M.: An enhanced agent based model of the immune system response. Cell Immunol. 244, 77–79 (2006)
8. Meyer-Hermann, M.E., Maini, P.K., Iber, D.: An analysis of B cell selection mechanisms in germinal centers. Math. Med. Biol. 23(3), 255–277 (2006)

9. Nerup, J., Mandrup-Poulsen, T., Helqvist, S., et al.: On the pathogenesis of IDDM. Diabetologia 37(Sup.2), S82–S89 (1994)

10. Van Nieuwenhuijze, A.E.M., van Lopik, T., Smeenk, R.J.T., Aarden, L.A.: Time between onset of apoptosis and release of nucleosomes from apoptotic cells: putative implications for systemic lupus erythematosus. Ann. Rheum. Dis. 62, 10–14 (2003)

11. Bluestone, J.A., Tang, Q., Sedwick, C.E.: T Regulatory Cells in Autoimmune Diabetes: Past Challenges, Future Prospects. J. Clin. Immunol. 28(6), 677–684 (2008)

12. Trudeau, J.D., Dutz, J.P., Arany, E., et al.: Neonatal β-cell Apoptosis: A Trigger for Autoimmune Diabetes? Diabetes 49(1), 1–7 (2000)

13. Marée, A.F.M., Kublik, R., Finegood, D.T., Edelstein-Keshet, L.: Modelling the onset of Type 1 diabetes: can impaired macrophage phagocytosis make the difference between health and disease? Philos. T. R. Soc. A. 364, 1267–1282 (2006)

14. Jo, J., Choi, M.Y., Koh, D.S.: Size Distribution of Mouse Langerhans Islets. Biophyis J. 93, 2655–2666 (2007)

15. Bonner-Weir, S.: β-cell Turnover: Its Assessment and Implications. Diabetes 50(Sup.1), 20–24 (2001)

16. The Jackson Laboratory (2001), Mouse Phenome Database, http://www.jax.org/phenome (Accessed 1st February 2010)

17. Klinke, D.J.: Extent of Beta Cell Destruction Is Important but Insufficient to Predict the Onset of Type 1 Diabetes Mellitus. PLoS ONE 3(1), e1374 (2008)

18. Bock, T., Pakkenberg, B., Buschard, K.: Genetic Background Determines the Size and Structure of the Endocrine Pancreas. Diabetes 54(1), 133–137 (2005)

19. Dor, Y., Brown, J., Martínez, O.I., Melton, D.A.: Adult pancreatic β-cells are formed by self-duplication rather than stem-cell differentiation. Nature 429, 41–46 (2004)

20. Meier, J.J., Butler, A.E., Saisho, Y., et al.: β-Cell Replication Is the Primary Mechanism Subserving the Postnatal Expansion of β-Cell Mass in Humans. Diabetes 57(6), 1584–1594 (2008)

21. Teta, M., Long, S.Y., Wartschow, L.M., et al.: Very Slow Turnover of β-Cells in Aged Adult Mice. Diabetes 54(9), 2557–2567 (2005)

22. Nir, T., Melton, D.A., Dor, Y.: Recovery from diabetes in mice by βcell regeneration. J. Clin. Invest. 117(9), 2553–2561 (2007)

23. Marée, A.F.M., Komba, M., Dyck, C., et al.: Quantifying macrophage defects in type 1 diabetes. J. Theor. Biol. 233, 533–551 (2005)

24. Scaglia, L., Cahill, C.J., Finegood, D.T., Bonner-Weir, S.: Apoptosis Participates in the Remodeling of the Endocrine Pancreas in the Neonatal Rat. Endocrinology 138(4), 1736–1741 (1997)

25. Montanya, E., Nacher, V., Biarnés, M., Soler, J.: Linear Correlation Between β-Cell Mass and Body Weight Throughout the Lifespan in Lewis Rats. Diabetes 49, 1341–1346 (2000)

State-of-the-Art Genetic Programming for Predicting Human Oral Bioavailability of Drugs

Sara Silva and Leonardo Vanneschi

Abstract. Being able to predict the human oral bioavailability for a potential new drug is extremely important for the drug discovery process. This problem has been addressed by several prediction tools, with Genetic Programming providing some of the best results ever achieved. In this paper we use the newest state-of-the-art developments of Genetic Programming, in particular the latest bloat control method, to find out exactly how much improvement we can achieve on this problem. We show examples of some actual solutions and discuss their quality from the practitioners' point of view, comparing them with previously published results. We identify some unexpected behaviors and discuss the way for further improving the practical usage of the Genetic Programming approach.

1 Introduction

The success of a drug treatment is strongly related with the ability of a molecule to reach its target in the patient's organism without inducing toxic effects. Human oral bioavailability (indicated with %F from now on) is the parameter that measures the percentage of the initial orally submitted drug dose that effectively reaches the systemic blood circulation after the passage from the liver. This parameter is particularly relevant, because the oral assumption is usually the preferred way for supplying drugs to patients and because it is a representative measure of the

Sara Silva
INESC-ID Lisboa, Rua Alves Redol 9, 1000-029 Lisboa, Portugal
e-mail: sara@kdbio.inesc-id.pt
and Center for Informatics and Systems of the University of Coimbra, Portugal

Leonardo Vanneschi
Department of Informatics, Systems and Communication (D.I.S.Co.),
University of Milano-Bicocca, Milan, Italy
e-mail: vanneschi@disco.unimib.it
and INESC-ID Lisboa, Portugal

M.P. Rocha et al. (Eds.): IWPACBB 2010, AISC 74, pp. 165–173, 2010.
springerlink.com

quantity of active principle that effectively can actuate its therapeutic effect. Being able to reliably predict the %F value for a potential new drug is outstandingly important, given that the majority of failures in compounds development from the early nineties to nowadays are due to a wrong prediction of this pharmacokinetic parameter during the drug discovery process [3, 4]. Most pharmacokinetics prediction tools reported in the literature belong to the category of Quantitative Structure-Activity Relationship (QSAR) models [11]. The goal of such models is to define a quantitative relationship between the structure of a molecule and its biological activity. See [1, 2] for an updated review of these prediction tools.

Genetic Programming (GP) [6] is a computational method aimed at learning computer programs that solve specific problems, given their high level specifications. Basically a search process, GP evolves populations of computer programs, using Darwinian evolution and Mendelian genetics as inspiration. It undoubtedly provides some of the most powerful and versatile problem solving algorithms developed so far, however its practical usage still poses a few challenges. Because GP uses a variable-length representation for the solutions, the programs are allowed to grow during the evolutionary process, and it usually happens that their size increases without a corresponding improvement of fitness. This is called *bloat*, a serious problem that can actually stagnate the evolutionary process, besides compromising the understandability of the provided solutions. From among the many theories explaining the emergence of bloat (reviews in [7, 8]), and the numerous methods attempting to prevent it (review in [7]), recent theoretical developments led to a new bloat control technique, called Operator Equalisation (OpEq) [9]. Although still recent and requiring improvements, OpEq has already proven to be more than just a bloat control method. It reveals novel evolutionary dynamics that allow, for the first time after more than 15 years of intense bloat research, a successful search without code growth.

GP has been used in pharmacokinetics in [5] for classifying molecules in terms of their %F, and also in [2, 1] for quantitatively predicting %F. These studies have shown that GP is a very promising approach, in most cases able to provide better solutions than the other machine learning methods studied. The problem of predicting %F has already been addressed by GP with OpEq [10], where the goal was to find whether the successfulness of OpEq in benchmark problems [9] would hold for a hard real-life regression problem. In the present paper we are only marginally interested in the evolutionary dynamics of the techniques. Instead, we focus our attention on the *actual* solutions that GP can provide to the problem of %F prediction, with and without OpEq. We put ourselves in the role of the practitioners and discuss the achievements from a practical point of view, comparing them with previously published results. We identify some unexpected behaviors and discuss the way for further improving the practical usage of GP.

In the next Section we describe the experiments performed in this study. Section 3 reports and discusses the results, and Section 4 concludes.

2   Experiments

We basically replicate the experiments of [10]. The techniques tested were a standard GP implementation (StdGP) and the two different versions of OpEq (DynOpEq and MutOpEq) described in [10]. Being GP a non-deterministic method (meaning that each GP run may produce a different result), a total of 30 runs were performed with each technique, to allow for statistic significance. All the runs used populations of 500 individuals allowed to evolve for 100 generations. The function set used to build the candidate solutions only contained the four arithmetic operators $+$, $-$, $\times$, and $/$. Fitness was calculated as the root mean squared error between outputs and targets, meaning that the lower the fitness, the better the individual. To create new individuals, crossover and mutation were used with probabilities 0.9 and 0.1, respectively. Survival from one generation to the other was always guaranteed to the best individual of the population. For the remaining parameters consult [10].

We have obtained a set of molecular structures and the corresponding %F values using the same data as in [13] and a public database of food and drug Administration (FDA) approved drugs and drug-like compounds [12]. The data has been gathered in a matrix composed by 359 rows and 242 columns. Each row is a vector of molecular descriptors values identifying a drug; each column represents a molecular descriptor, except the last one, that contains the known values of %F[1]. Training and test sets have been obtained by randomly splitting the dataset: at each GP run, 70% of the molecules have been randomly selected with uniform probability and inserted into the training set, while the remaining 30% form the test set. For more details consult [10].

All the experiments were performed using a modified version of GPLAB[2]. Statistical significance of the null hypothesis of no difference was determined with pairwise Kruskal-Wallis non-parametric ANOVAs at $p = 0.01$.

3   Results and Discussion

In the following subsections we use the term *training fitness* to designate the fitness measured on the training set, and the term *test fitness* to designate the fitness measured on the test set (by the same individual, unless otherwise indicated). Although debatable from a practitioners' point of view, we use the term *best solution* to designate the solution represented by the individual with the best training fitness. We use the term *length* of a solution to designate the total number of variables, constants and operators in the final expression.

---

[1] This dataset, and a lookup table with descriptor acronyms, can be downloaded from, respectively: http://personal.disco.unimib.it/Vanneschi/bioavailability.txt
http://personal.disco.unimib.it/Vanneschi/bioavailability_lookup.txt

[2] GPLAB – A Genetic Programming Toolbox for MATLAB, freely available at:
http://gplab.sourceforge.net

Fitness and Length of Solutions. Figure 1 contains two boxplots. The first one (a) refers to the best training fitness achieved on the last generation. The second (b) shows the fitness achieved by the same individual when measured on the test set. All the differences are statistically significant on the training set, where DynOpEq is the technique that reaches better fitness, followed by StdGP, and finally MutOpEq. Remarkably enough, no significant differences are observed on the test set. However, despite all techniques having similar generalization ability, DynOpEq exhibits a higher variability of test fitness, with two outliers falling outside the plot (as hinted by the magnitude of the mean value). StdGP and MutOpEq exhibit a more constrained behavior, although MutOpEq also has one outlier outside the plot.

Figure 2 contains two plots. The first (a) is an unconventional plot that shows the evolution of the training fitness of the best individual plotted against its length, median of 30 runs. There is an implicit downwards timeline along the fitness axis. Depending on how fast the fitness improves with the increase of program length, the lines in the plot may point downward (south), or they may point to the right (east). Lines pointing south represent a rapidly improving fitness with little or no code growth. Lines pointing east represent a slowly improving fitness with strong code growth. Lines pointing southwest (bottom left) represent improvements in fitness along with a reduction of program length. We want our lines to point as south (and west) as possible. As can be seen, both OpEq techniques point down, while StdGP points mostly right. At the end of the run there is a significant difference between all the techniques, with StdGP producing the longest solutions, and MutOpEq producing the shortest. From [10] we already knew that the lines referring to *average* solution length behaved like this. Knowing that the *best* solution length follows the same trend reveals that OpEq not only allows a better search for shorter solutions, but it actually *finds* them. The second plot (b) shows the evolution of the best solution's length along the generations, median of 30 runs. Here we can once again observe the striking difference between StdGP, with its steadily growing solution

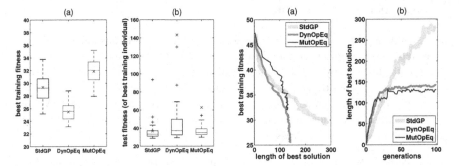

Fig. 1 Boxplots of (a) best training fitness and (b) test fitness of best training individual. The + represent outliers, and × marks the mean

Fig. 2 Evolution of (a) training fitness versus length of best solution and (b) length of best solution per generation

length, and the OpEq techniques with their stabilizing solution length from early in the run.

Examples of Solutions. So far, the previous results have shown that the solutions obtained by GP with OpEq are usually shorter than the ones obtained without OpEq. Now we look at some actual solutions obtained during our experiments. Figure 3 shows some of the shortest best solutions provided in the last generation, by the three different techniques. These expressions were chosen for being the shortest *after* removing the redundant code and performing a symbolic simplification with MATLAB, and not necessarily the shortest in the raw format returned by GP. From among the 30 runs, we show three solutions (1,2,3) provided by MutOpEq, one (4) by DynOpEq, and also two of the shortest best solutions (5,6) provided by StdGP (and an extra solution provided by DynOpEq, explained later). For each of them we indicate the technique and run that achieved it, the original length of the raw expression, and the fitness measured on the training and test datasets.

Comparison with Previous Results. Previous studies on the prediction of %F [1, 2] have compared the results of different machine learning techniques, with and without prior feature selection, and concluded that some GP techniques can provide the best results in terms of the fitness measured in the test set. A direct comparison between these previous results and the present study is not possible, as the experimental setup was not the same, and the results were reported differently. Nevertheless, they are our best term of comparison, so we perform a very light comparison.

The previous studies report the solution that performed better in the test set, from among *all* the individuals ever created with a standard GP technique, to achieve a test fitness of 30.13. The best test fitness achieved by any of the other GP techniques used in that study was 26.01, using linear scaling and ephemeral random constants [1, 2]. The raw expression of this solution had length 201. We were curious to know whether, by looking among the fitness values measured on the *test* set, we would also find a solution of similar quality. Without doing an exhaustive search, we easily found a solution with fitness values of 26.43 and 25.62 on the training and test sets, respectively. Its expression is the last one (7) in Figure 3.

Further Analysis. From the practitioners' point of view, the last expression of Figure 3 is visually more complex than the other expressions from the OpEq techniques, but its good and similar values of training and test fitness make it very appealing in terms of generalization ability. This solution was created by DynOpEq in run 25, and we found it in rank 180 among the 500 individuals of the last generation. The first place in this ranking consisted of a solution with fitness value of 87.62 on the test set, one of the outliers shown in Figure 1(b). At first, this seemed like a typical bad case of overfitting, but a closer look revealed an unexpected behavior. At the end of generation 52 the best solution had fitness values of 27.33 and 26.76 in the training and test set, respectively. In the next generation the test fitness of the best individual had "jumped" to 89.07, and it remained around this value for the rest of the run. In the last generation, the individual ranked 180 is just one of the many

MutOpEq, run 4, original length 115, training fitness 32.99, test fitness 34.41:

$$\frac{x_{156}\,(x_{30}+x_{132})}{x_{135}} + 2x_{182} + 3x_{171} + \frac{x_{18}\,x_{64}}{x_{156}} + 2x_{222} + 2\frac{x_{156}\,x_{24}}{x_{135}} + 3\frac{x_{30}}{x_{18}} + \frac{x_{171}}{x_{184}} + \frac{x_{18}+x_{100}}{x_{92}\,x_{186}} + \frac{x_{180}}{x_{156}} - x_{80} +$$

$$+2x_{134} - 2x_{91} + \frac{x_{30}+x_{132}}{x_{24}} - x_{171}\left(x_{64} + \frac{x_{180}}{x_{171}}\right)x_{184}^{-1}x_{73}^{-1}x_{222}^{-1} - \frac{x_{30}+x_{132}}{x_{184}\,x_{228}} + x_{230} \quad (1)$$

MutOpEq, run 21, original length 75, training fitness 27.96, test fitness 32.93:

$$\left[2x_2 + \frac{x_2}{x_{30}} + x_{58} - (x_6 - x_{35})x_{91} + x_{231}\right]x_{30}^{-1}\left(\frac{x_2+x_{231}}{x_{200}} + x_{143}\right)^{-1} + \frac{x_{30}\,x_{45}}{x_{149}} + \frac{x_{205}}{x_{143}} +$$

$$+\left(x_{182} + \frac{x_{116}}{x_{131}}\right)x_{219}^{-1} - x_{164}\,(x_{160} - x_{219})\left[x_{200} + x_{219}^{-1}\left(x_{110} - \frac{x_2}{x_{30}} - x_{58}\right)^{-1}\right]^{-1} \quad (2)$$

MutOpEq 27, original length 133, training fitness 33.43, test fitness 30.50:

$$x_{231} + x_{227} + \frac{x_{125}}{x_{124}\,x_{227}} + x_{91}\,x_{224} + 2\frac{x_{191}}{x_{218}} + 2\frac{x_{151}\,x_{222}}{x_{142}\,x_{227}} + 3\frac{x_{201}}{x_{91}\,x_{170}} + \frac{x_{142}\,x_{164}\,x_{97}}{x_{222}\,x_{224}} - 5x_{124} + \frac{x_{199}}{x_{32}\,x_{227}} +$$

$$+\frac{x_{164}}{x_{227}\,x_{199}} + x_{164}\,x_{222}\,x_{225} - x_{91} + x_5 - x_{12} - x_{18} + x_{191}^{-1} + 2x_{199}^{-1} + \frac{x_{106}}{x_{97}} \quad (3)$$

DynOpEq 14, original length 129, training fitness 27.72, test fitness 35.76:

$$2x_{38} - \left(8x_{94} + 2x_{18} + 23x_{28} + \frac{x_{101}}{x_{221}^2} + \frac{x_{182}}{x_{221}} - 2x_{232}\,x_{41} + \frac{x_{38}}{x_{221}} - x_{232} - x_{26}^2\,x_{94} + x_{41} + \frac{x_{28}}{x_{221}}\right)$$

$$(x_{52} - x_{189} - x_{28})^{-1} + 2x_{26} - 3x_{18} + 2x_{50} + x_{230} \quad (4)$$

StdGP, run 8, original length 207, training fitness 27.46, test fitness 30.77:

$$x_{17} + \left[1 + 2\frac{x_{231}}{x_{212}\,x_{45}} + (2x_4 + x_{231})\left(x_9 - x_{61} - \frac{x_{82}}{x_4}\right)(x_4 + x_{30})^{-1} + x_5\,x_{18}\left(\frac{x_5}{x_{212}} + 2x_{17}\right)^{-1}x_{216}^{-1}x_{45}^{-1} + (x_4 + x_{30})\right.$$

$$\left(x_9 - x_{61} - \frac{x_{82}}{x_{238}}\right)^{-1} + x_{212} + \frac{x_{17}}{x_{212}} + 10x_{17} + 3x_{30} + \frac{x_5}{x_{212}} + \frac{x_{17}}{x_{18}} + \left(\frac{x_{17}}{x_{18}} + x_{30}\right)(x_9 - x_{118} - x_{216}^{-1})^{-1} + \frac{x_{17}}{x_4 + x_{231}} +$$

$$+\frac{2x_4 + x_{30}}{x_{216}} + \frac{x_4 + 4x_{231} + 4x_{17} + 2x_{37} + x_{30}}{x_{18}} + 2x_{211} + 2\frac{x_{17}}{x_{216}} + 5x_4 + 2x_5 + 5x_{37} + 4x_{231}\right]x_{30}^{-1} - x_4 \quad (5)$$

StdGP, run 12, original length 135, training fitness 28.50, test fitness 30.51:

$$x_{45} - \left[x_3 + x_{64} - x_{46} - \left(\frac{x_{17}}{x_{37}\,x_{131}} - x_{30} - 3x_{56} + x_{17}\right)\left(x_{37} - \left((x_{56} - x_{17})\left(\frac{x_{152}}{x_{37}} - x_{30} - x_{56} + x_{17}\right)^{-1} - x_{94} - x_{231}\right)\right.\right.$$

$$((x_{131} - x_{17})x_{200} - x_{17})^{-1} - x_{30} - x_{76} + x_{56}\Big)^{-1}\Bigg]\left[\left(\frac{x_{17}}{x_{30} + x_{76} - x_{56}} - x_{17}\right)x_{37}^{-1} - x_{94} - x_{231}\right][(x_{152} - x_3)x_{200} - x_{17}]^{-1}$$

$$\left[\frac{x_{37}}{x_{231} - x_{56}\,(x_{64} - x_{46})x_{239} - x_{17}} - x_{239} + x_{121} - x_{30}\right]^{-1} \quad (6)$$

DynOpEq, run 25, original length 153, training fitness 26.43, test fitness 25.62:
(individual ranked 180 in the last generation)

$$\left[x_{37} + x_{45}\left(x_{161} + x_{89} - \frac{x_{106}\,(x_{133} + x_{231} - x_{177})}{x_{89}\,x_{133}}\right)^{-1}(x_{161} + x_{23})^{-1} + \frac{(x_{218} + x_{89} - x_{210})x_{45}}{x_{232}\,(x_{161} + x_{23})} +\right.$$

$$+x_{231}\,x_{210} + x_{89} - x_{127}\,x_{89}^{-1}\left(\left(x_{133} + x_{231} - x_{177} + x_{28} + \frac{x_{89}}{x_{42}} - \frac{x_3}{x_{29}}\right)x_{29}^{-1} + x_{232} + x_{45}\right)^{-1} -$$

$$-x_{127} - 2x_{85} - x_{133} - x_{122}]x_{19}^{-1}x_{28}^{-1}\left(\frac{x_{122}}{x_{89}} + x_{89} - x_{210}\right)^{-1} + \frac{x_1}{x_{29}} + \frac{x_3}{x_{231} - x_{89} - x_{177}} +$$

$$+x_1\left[\frac{x_{85} - x_{29}}{x_{85}} + x_{231} - x_{29} + \frac{x_{19}\,x_{51}\,(x_{127} + x_{231} - x_{89} - 2x_{177} + x_{28})x_1}{x_{89}\,(x_{45} + x_{127})x_{29}} + \frac{x_{106}}{x_{42}}\right]^{-1} \quad (7)$$

Fig. 3 Examples of solutions provided by the different GP techniques. Lookup table with descriptor acronyms available at http://personal.disco.unimib.it/Vanneschi/bioavailability_lookup.txt

remnants of the "pre-jump" phase of the evolution. We soon realized this was not an isolated event, and identified many cases of extreme variability of the test fitness values along the runs. When the 30 runs are regarded in median terms these variations go unnoticed, but this is certainly a subject for future investigation. It may not be the typical overfitting behavior, and can be caused by possible errors in the data, but the implementation of a method to avoid overfitting, even a simple one like the early stopping of learning, could prevent good solutions from being lost during the evolution. Overfitting has not yet been intensively studied in GP, but it is another problem that needs to be solved in order to allow its practical and reliable usage.

Feature Selection. Many machine learning techniques rely on prior feature selection to reduce the dimensionality of the search space, but GP is known for performing automatic feature selection during the search. The process that allows this is simple: In this particular problem, GP searches the space of all arithmetic expressions of 241 variables. This space contains expressions that use all the variables, and expressions that use a strict subset of them. There is no reason why an expression using less than the 241 variables cannot have better fitness than an expression using all the variables. Given that fitness is the only principle that guides the evolutionary search, GP may return a lower dimensional expression as the best solution found, thus performing automatic feature selection. So, besides seldom returning the same solution, GP also rarely chooses the same set of features in two different runs, particularly in complex high-dimensional problems like the one we are studying. Promoting the creation of alternative solutions for the same problem has been the subject of a number of studies, as it potentially enhances creativity and innovation, desirable properties in many application domains [14].

Table 1 lists the features used to construct each of the solutions shown in Figure 3. Not many features appear in more than one solution (indicated in bold). This small set of example solutions is representative of the entire set of solutions generated for this problem. From the 241 features available to predict %F, all the techniques select more or less the same number of features, but not the same ones. We could not identify a core of preferred features, not even a single feature that is always selected

Table 1 Features selected by the solutions of Figure 3. Lookup table with descriptor acronyms available at http://personal.disco.unimib.it/Vanneschi/bioavailability_lookup.txt

Solution	Features list (in bold, the ones that appear more than once)
(1)	$x_{18}, x_{24}, \mathbf{x_{30}}, \mathbf{x_{64}}, x_{73}, \mathbf{x_{80}}, \mathbf{x_{91}}, x_{92}, x_{100}, x_{132}, x_{134}, x_{135}, x_{156}, x_{171}, x_{180}, \mathbf{x_{182}}, x_{184}, x_{186}, \mathbf{x_{222}}, x_{228}, \mathbf{x_{230}}$
(2)	$x_2, x_6, \mathbf{x_{30}}, x_{35}, \mathbf{x_{45}}, x_{58}, \mathbf{x_{91}}, x_{110}, x_{116}, \mathbf{x_{131}}, x_{143}, x_{149}, x_{160}, \mathbf{x_{164}}, \mathbf{x_{182}}, \mathbf{x_{200}}, x_{205}, x_{219}, \mathbf{x_{231}}$
(3)	$x_5, x_{12}, \mathbf{x_{18}}, x_{32}, \mathbf{x_{91}}, x_{97}, \mathbf{x_{106}}, x_{124}, x_{125}, x_{142}, x_{151}, \mathbf{x_{164}}, x_{170}, x_{191}, x_{199}, x_{201}, \mathbf{x_{218}}, \mathbf{x_{222}}, x_{224}, x_{225}, x_{227}, \mathbf{x_{231}}$
(4)	$x_{18}, x_{26}, \mathbf{x_{28}}, x_{38}, x_{41}, x_{50}, x_{52}, \mathbf{x_{94}}, x_{101}, \mathbf{x_{182}}, x_{189}, x_{221}, \mathbf{x_{230}}, \mathbf{x_{232}}$
(5)	$x_4, \mathbf{x_5}, x_9, \mathbf{x_{17}}, \mathbf{x_{18}}, \mathbf{x_{30}}, \mathbf{x_{37}}, \mathbf{x_{45}}, x_{61}, x_{82}, x_{118}, x_{211}, x_{212}, x_{216}, \mathbf{x_{231}}, x_{238}$
(6)	$x_3, \mathbf{x_{17}}, \mathbf{x_{30}}, \mathbf{x_{37}}, \mathbf{x_{45}}, x_{46}, x_{56}, \mathbf{x_{64}}, x_{76}, \mathbf{x_{94}}, x_{121}, \mathbf{x_{131}}, x_{152}, \mathbf{x_{200}}, \mathbf{x_{231}}, x_{239}$
(7)	$x_1, \mathbf{x_3}, x_{19}, x_{23}, \mathbf{x_{28}}, x_{29}, \mathbf{x_{37}}, x_{42}, \mathbf{x_{45}}, x_{51}, x_{85}, x_{89}, \mathbf{x_{106}}, x_{122}, x_{127}, x_{133}, x_{161}, x_{177}, x_{210}, \mathbf{x_{218}}, \mathbf{x_{231}}, \mathbf{x_{232}}$

by either of the techniques. There are only two features, x_{230} and x_{231} (respectively N_IoAcAt and N_IoBaAt, two descriptors related to ionization in water), that stand out for being selected more often than the remaining 239.

4 Conclusions

We have used the newest state-of-the-art developments of GP for predicting the human oral bioavailability of medical drugs from a set of molecular descriptors. We have shown that the latest bloat control method allows the production of much shorter solutions, and identified some unexpected behaviors loosely related to overfitting, whose future resolution will allow a further improvement of results.

Acknowledgements. The authors acknowledge project "EnviGP – Improving Genetic Programming for the Environment and Other Applications" (PTDC/EIA-CCO/103363/2008) from Fundação para a Ciência e a Tecnologia, Portugal.

References

1. Archetti, F., Lanzeni, S., Messina, E., Vanneschi, L.: Genetic programming for human oral bioavailability of drugs. In: Cattolico, M. (ed.) Proceedings of GECCO-2006, pp. 255–262 (2006)
2. Archetti, F., Messina, E., Lanzeni, S., Vanneschi, L.: Genetic programming for computational pharmacokinetics in drug discovery and development. Genetic Programming and Evolvable Machines 8(4), 17–26 (2007)
3. Kennedy, T.: Managing the drug discovery/development interface. Drug Discovery Today 2, 436–444 (1997)
4. Kola, I., Landis, J.: Can the pharmaceutical industry reduce attrition rates? Nature Reviews Drug Discovery 3, 711–716 (2004)
5. Langdon, W.B., Barrett, S.J.: Genetic Programming in data mining for drug discovery. In: Evolutionary computing in data mining, pp. 211–235 (2004)
6. Poli, R., Langdon, W.B., McPhee, N.F.: A field guide to genetic programming. (With contributions by J. R. Koza) (2008), Published via, http://lulu.com and freely available at, http://www.gp-field-guide.org.uk
7. Silva, S.: Controlling bloat: individual and population based approaches in genetic programming. PhD thesis, Departamento de Engenharia Informatica, Universidade de Coimbra (2008)
8. Silva, S., Costa, E.: Dynamic limits for bloat control in genetic programming and a review of past and current bloat theories. Genetic Programming and Evolvable Machines 10(2), 141–179 (2009)
9. Silva, S., Dignum, S.: Extending operator equalisation: Fitness based self adaptive length distribution for bloat free GP. In: Vanneschi, L., Gustafson, S., Moraglio, A., De Falco, I., Ebner, M. (eds.) EuroGP 2009. LNCS, vol. 5481, pp. 159–170. Springer, Heidelberg (2009)
10. Silva, S., Vanneschi, L.: Operator Equalisation, Bloat and Overfitting - A Study on Human Oral Bioavailability Prediction. In: Rothlauf, F., et al. (eds.) Proceedings of GECCO-2009, pp. 1115–1122. ACM Press, New York (2009)

11. Van de Waterbeemd, H., Rose, S.: The Practice of Medicinal Chemistry. In: Wermuth, L.G. (ed.), 2nd edn., pp. 1367–1385. Academic Press, London (2003)
12. Wishart, S.D., Knox, C., Guo, A.C., Shrivastava, S., Hassanali, M., Stothard, P., Chang, Z., Woolsey, J.: DrugBank: a comprehensive resource for in silico drug discovery and exploration. Nucleic Acids Research 34 (2006), doi:10.1093/nar/gkj067
13. Yoshida, F., Topliss, J.G.: QSAR model for drug human oral bioavailability. Journal of Medicinal Chemistry 43, 2575–2585 (2000)
14. Zechman, E.M., Ranjithan, S.R.: Multipopulation Cooperative Coevolutionary Programming (MCCP) to Enhance Design Innovation. In: Beyer, H.G., et al. (eds.) Proceedings of GECCO-2005, pp. 1641–1648. ACM Press, New York (2005)

Pharmacophore-Based Screening as a Clue for the Discovery of New P-Glycoprotein Inhibitors

Andreia Palmeira, Freddy Rodrigues, Emília Sousa, Madalena Pinto, M. Helena Vasconcelos, and Miguel X. Fernandes[*]

Abstract. The multidrug resistance (MDR) phenotype exhibited by cancer cells is believed to be hampering successful chemotherapy treatment in cancer patients. A group of ABC drug transporters which particularly include P-glycoprotein (Pgp) contribute to this phenotype. Thus, there is a need to anticipate whether drug candidates are possible Pgp substrates or noncompetitive inhibitors. Therefore, a pharmacophore model was created based on known Pgp inhibitors and it was used to screen a database of commercial compounds. After the screening, twenty-one candidate compounds were selected and their influence in the intracellular accumulation of Pgp substrate Rhodamine-123 (Rh123) was investigated by flow cytometry. Eleven compounds were found to significantly increase the accumulation of Rh123, four were found to decrease and six showed only a slight effect on the accumulation of Rh123. Furthermore, the competitive/non-competitive mechanism for the most promising compounds was determined by a luminescence Pgp's ATPase assay. Based on the cytometry results, a new pharmacophore was created for the Pgp inhibitory activity. The overall results provide important clues on how to proceed towards the discovery of Pgp inhibitors and which type of molecules merit further analysis.

Andreia Palmeira · Emília Sousa · Madalena Pinto
Laboratory of Organic Chemistry, Faculty of Pharmacy, University of Porto,
Rua Anibal Cunha 164, 4050-047, Porto

Andreia Palmeira · Emília Sousa · Madalena Pinto
Research Center of Medicinal Chemistry (CEQUIMED-UP), University of Porto, Portugal,
Rua Anibal Cunha 164, 4050-047, Porto

Andreia Palmeira · M. Helena Vasconcelos
Cancer Biology Group, IPATIMUP - Institute of Molecular Pathology and Immunology of the University of Porto, Portugal, Rua Dr. Roberto Frias, s/n, 4200-465 Porto

Freddy Rodrigues · Miguel X. Fernandes
Centro de Química da Madeira, Centro de Competência de Ciências Exactas e Engenharias,
Universidade da Madeira, Campus da Penteada, 9000-390, Funchal, Portugal
e-mail: mfx@uma.pt

M. Helena Vasconcelos
Laboratory of Microbiology, Faculty of Pharmacy, University of Porto, Portugal,
Rua Anibal Cunha 164, 4050-047, Porto

[*] Corresponding author.

M.P. Rocha et al. (Eds.): IWPACBB 2010, AISC 74, pp. 175–180, 2010.
springerlink.com   © Springer-Verlag Berlin Heidelberg 2010

1 Introduction

Resistance to chemotherapy represents a major obstacle in the treatment of cancer. Multidrug resistance (MDR) can be broadly defined as a phenomenon by which tumor cells *in vivo* and cultured cells *in vitro* show simultaneous resistance to a variety of structurally and functionally dissimilar cytotoxic and xenobiotic compounds (Higgins 2007). P-glycoprotein (Pgp), a 170-KDa plasma membrane protein, represents one of the best characterized barriers to chemotherapeutic treatment in cancer. Pgp actively transports to the extracellular space structurally unrelated compounds, conferring the MDR phenotype in cancer. A logical step to reverse MDR phenotype is finding molecules that can directly block the activity of Pgp (Lehne 2000). However, this is a difficult process since the available Pgp structural data has only low-to-medium resolution. Therefore the most suited approach for designing new compounds is ligand-based design (Pajeva et al. 2004).

A direct way to analyse Pgp activity is to determine the mean fluorescence intensity of cells jointly exposed to Rhodamine123 (Rh123), a known Pgp substrate, and the potential Pgp inhibitor (Goda et al. 2009). The approach is simple and fast, though indirect. Nonetheless, compounds that enhance the retention of the probe are scored as Pgp ligands, because they interfere with the extrusion of the marker. In order to determine how a ligand interacts with Pgp, the measurement of its effect on the rate of Pgp's ATP hydrolysis allows to discriminate between noncompetitive and competitive inhibitors which are themselves substrates for transport (Matsunaga et al. 2006).

Herein, we investigated the effect of twenty-one compounds in the accumulation of Rh123 in a leukemia cell line overexpressing Pgp, K562Dox, applying a pharmacophore model. The Pgp's ATP hydrolysis assay was used to distinguish between noncompetitive and competitive inhibitors (substrates). A 3D-QSAR pharmacophore model was also created according to the results obtained in the Rh123 accumulation assay.

2 Methods

2.1 Pharmacophore Modeling and Virtual Screening of a Database of Compounds

A pharmacophore is the 3D arrangement with the minimal molecular features needed for a compound to show biological activity and is given as a set of distance restraints between features. To create the pharmacophore model we started with 26 Pgp known inhibitors from the flavonoid family. The inhibitors structural data were uploaded in the PharmaGist webserver (Schneidman-Duhovny et al. 2008). After getting the pharmacophore from Pharmagist we used it to screen the database DrugBank (Wishart et al. 2006). To do so, we first edited the database files using small scripts to remove counter-ions, neutralization of partial charges and removal of solvent molecules. Additionally, we converted all structures to 3D using the program Corina (Molecular Networks GmbH, Erlangen, Germany). After preparing the database we run the program VLifeMDS (VLife Ltd., Pune,

India) to screen virtually the database using the pharmacophore model generated earlier with distance tolerances of 20%. After the screening we discarded the hits with a RMSD, to the pharmacophore model, greater than 0,1Å and also discarded the hits that did not comply with the Lipinski rules (Lipinski 2000).

2.2 Flow Cytometry Using Rhodamine 123

K562 (human chronic myelogenous leukaemia, erythroblastic) and K562Dox (derived from K562, overexpesses Pgp) cell lines (Lima et al. 2007) in exponential growth were used in cytometry assays. K562Dox (5×10^6 cells/mL) were incubated for 1 h in the presence of 10 or 20μM of the test compounds, and with 1μM Rh123. K562Dox and K562 alone as well as K562Dox in the presence of Verapamil, Mibefradil and Quinidine (10 and 20 μM), known Pgp inhibitors, were used as controls. After the incubation time, cells were washed twice, ressuspended in ice cold PBS and kept at 4°C in the dark until analysis in the flow cytometer. At least 20 000 cells per sample were counted and analyzed by flow cytometry (Epics XL-MCL, Coulter), and the amount of fluorescence was plotted as a histogram of FL1. Data acquisition was performed using WINMDI (version 2.9) to determine median fluorescence intensity values.

2.3 Determination of ATPase Activity

The ATPase activity of Pgp was determined using the luminescent ATP detection kit (Pgp-Glo Assay Kit, Promega) (Dongping M et al. 2007), using sodium vanadate (Na_3VO_4) and Verapamil as controls. Test compounds at 200μM in buffer solution were incubated with 0.5 mg/mL Pgp and 5 mM MgATP at 37°C for 40 min, and the remaining ATP was detected as a luciferase-generated luminescent signal in a luminometer.

2.4 Pharmacophore Hypothesis

A 3D-QSAR pharmacophore model was also created using HypoGen module of Catalyst program (Accelrys v2.1) (Guner et al. 2004) according to the results obtained in the accumulation of Rh123 assay for 15 Pgp inhibitors. Conformer generation was carried out with the "Best" algorithm, the feature groups selected were hydrogen bond (Hb)-donor and -acceptor, hydrophobic, positive and negative ionisable and remaining default parameters. Validation of the pharmacophore, prior to the next cycle of experimental determinations, was performed using an enrichment test with 1000 decoy molecules from National Cancer Institute (NCI) with similar characteristics (no. atoms, MW, no. Hb-donnor and Hb-acceptors) and twelve known Pgp inhibitors.

3 Results and Discussion

There were 3 common features present in all 26 inhibitors, 2 aromatic rings and 1 H-bond acceptor and they can be depicted as shown in Figure 1. After the procedures described to clean the database, we used 4825 structures from DrugBank

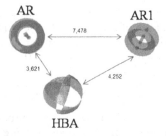

Fig. 1 Pharmacophore hypothesis used for virtual screening of potential Pgp inhibitors

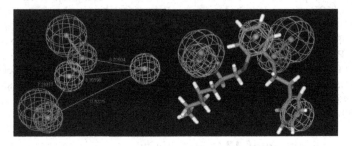

Fig. 2 Putative pharmacophore for Pgp inhibitors. Green= Hb-acceptor, blue= hydrophobic region; inter-center distances are represented (top). Pharmacophore and Propafenone (botton).

Fig. 3 Enrichment test for pharmacophore model

which were screened to determine their compliance with the pharmacophore model. The hits retrieved follow a normal distribution with $R^2 = 0,828$ and we found 167 structures that comply with the pharmacophore model with a RMSD of less than 0,1Å. Of these, 20 commercially available molecules were purchased for *in vitro* testing.

Since K562Dox cell line overexpresses Pgp, differences in the accumulation pattern between the several treatments should be related to modulation of this pump. According to the cytometry assay we found that Verapamil, Quinidine and Mibefradil (known Pgp inhibitors) as well as eleven of the new investigated compounds are increasing the intracellular accumulation of Pgp substrate Rh123. Particularly, Propafenone, Azelastine, Amoxapine, and Loxapine showed an effect similar to that elicited by a known Pgp inhibitor, Quinidine. In contrast, Blebbistatin, Coelenteramide, Indirubin and Leflunomide showed an effect compatible with Pgp activation; only a slight effect in the accumulation rate of Rh123 was observed for the other selected compounds.

To elucidate which was the mechanism of action of the most promising compounds, their effect on the Pgp ATPase activity was investigated. Results showed that Econazole, Amoxapine, Loxapine, Bicalutamide, Zomepirac and Tioconazole are noncompetitive inhibitors of Pgp, blocking the ATPase activity. On the other hand, Propafenone, Hycanthone, Cyclic Pifithryn-α, Diltiazem, Azelastine and Prazosin are like Verapamil, competitive inhibitors of Pgp, stimulating the ATP hydrolysis and being themselves transported by the pump.

Based on the the cytometry assay results, a new refined pharmacophore model was constructed (Figure 2). This pharmacophore included two hydrophobic (HA) regions and two Hb-acceptor (HBA) features which were intercalated with each other. An enrichment test to validate the pharmacophore, prior to validation in the next cycle of screening and experimental determinations, is represented in Figure 3 and shows enrichment rates of 10-fold over random retrieval of compounds from a pool of unknown molecules. This new pharmacophore retains features that explain better the differences between experimental activities. It has one more feature than the initial one (Figure 1) but the similarities are clear. These results will guide the investigation of other molecules, namely new compounds obtained by synthesis, in order to validate the pharmacophore, and the accuracy in the prediction of activity.

Acknowledgements

FCT (I&D, n° 226/2003 and I&D, n° 4040/07), FEDER, POCI, for financial support.

References

Dongping, M.: Identify P-glycoprotein Substrates and Inhibitors with the Rapid, HTS Pgp-Glo™ Assay System (2007),
http://www.promega.com/pnotes/96/15080_11/15080_11.pdf
(Accessed February 12, 2010)
Goda, K., Bacso, Z., et al.: Multidrug resistance through the spectacle of P-glycoprotein. Curr. Cancer Drug Targets 9, 281–297 (2009)
Guner, O., Clement, O., et al.: Pharmacophore modeling and three dimensional database searching for drug design using catalyst: recent advances. Curr. Med. Chem. 11, 2991–3005 (2004)
Higgins, C.F.: Multiple molecular mechanisms for multidrug resistance transporters. Nature 446, 749–757 (2007)
Lehne, G.: P-glycoprotein as a drug target in the treatment of multidrug resistant cancer. Curr. Drug Targets 1, 85–99 (2000)
Lima, R.T., Guimarães, J.E., et al.: Overcoming K562Dox resistance to STI571 (Gleevec) by downregulation of P-gp expression using siRNAs. Cancer Ther. 5, 67–76 (2007)
Lipinski, C.A.: Drug-like properties and the causes of poor solubility and poor permeability. J Pharmacol Toxicol Methods 44, 235–249 (2000)

Matsunaga, T., Kose, E., et al.: Determination of p-glycoprotein ATPase activity using luciferase. Biol. Pharm. Bull. 29, 560–564 (2006)

Pajeva, I.K., Globisch, C., et al.: Structure-function relationships of multidrug resistance P-glycoprotein. J. Med. Chem. 47, 2523–2533 (2004)

Schneidman-Duhovny, D., Dror, O., et al.: PharmaGist: a webserver for ligand-based pharmacophore detection. Nucleic Acids Res. 228, W223–W228 (2008)

Wishart, D.S., Knox, C., et al.: DrugBank: a comprehensive resource for in silico drug discovery and exploration. Nucleic Acids Res. 672, D668–D672 (2006)

e-BiMotif: Combining Sequence Alignment and Biclustering to Unravel Structured Motifs

Joana P. Gonçalves and Sara C. Madeira

Abstract. Transcription factors control transcription by binding to specific sites in the DNA sequences of the target genes, which can be modeled by structured motifs. In this paper, we propose *e*-BiMotif, a combination of both sequence alignment and a biclustering approach relying on efficient string matching techniques based on suffix trees to unravel all approximately conserved blocks (structured motifs) while straightforwardly disregarding non-conserved regions in-between. Since the length of conserved regions is usually easier to estimate than that of non-conserved regions separating the binding sites, ignoring the width of non-conserved regions is an advantage of the proposed method over other motif finders.

1 Introduction

Transcription factors (TFs) are key elements of regulatory mechanisms binding to specific sites in the DNA of protein coding genes to enhance or inhibit their transcription into mRNA. Binding sites are short stretches with 5 to 25 nucleotides long usually located within non-coding parts of DNA in the so-called promoter regions. Promoter regions can be modeled as structured motifs, that is, sets of conserved sequences, or motifs, separated by non-conserved regions of unspecified length. It is known that genes regulated by common TFs share identical binding sites. In this context, similar promoter regions can provide important insights into functionally relatedness of genes and their corresponding regulatory mechanisms. Identifying structured motifs is a challenging problem. On one hand, we do not know whether a given set of sequences is regulated by the same TFs or not. On the other hand, prior information on the composition of the motif is usually unavailable.

Joana P. Gonçalves · Sara C. Madeira

KDBIO group, INESC-ID, Lisbon, and Instituto Superior Técnico,

Technical University of Lisbon, Portugal

e-mail: `jpg@kdbio.inesc-id.pt, smadeira@kdbio.inesc-id.pt`

M.P. Rocha et al. (Eds.): IWPACBB 2010, AISC 74, pp. 181–191, 2010.

Methods for identifying structured motifs in sequences follow either probabilistic or combinatorial approaches. Most probabilistic strategies rely on iterative Expectation-Maximization (EM) or Gibbs sampling approaches [1, 8], where convergence is not guaranteed to lead to a global maximum and noisy data is an issue.

Combinatorial methods typically enumerate all possible motifs and search for their occurrences in the sequences using heuristics to reduce the exponential search space [2, 3, 5, 6, 10]. However, they are not able to discriminate relevant motifs from potentially numerous random motifs matching the generated models. Furthermore, they require large number of parameters. This is a major limitation when little is known about the configuration of the motifs and restricts the motifs to be found.

We present e-BiMotif, a method to identify structured motifs, ignoring the width of non-conserved regions separating the binding sites. This presents an advantage over other motif finders, since the length of conserved regions is usually easier to estimate. Following the approach of Wang et al. [9], e-BiMotif combines sequence alignment, to reveal conserved regions within sequences, with biclustering to further group sequences with similar motif structures.

This paper is outlined as follows. Section 2 describes alternative methods for structured motif identification. Section 3 presents our approach. Finally, Section 4 outlines conclusions and future work.

2 Related Work

SMILE [5] relies on a traversal with backtracking of a generalized suffix tree to spell all occurrences of simple motifs for a structured motif with gaps. The search for each motif is guided by a virtual lexicographic trie containing all possible motifs in a range of lengths. Its time complexity is exponential in the number of gaps [2].

Co-Bind [8] models structures of two motifs using Position Weight Matrices (PWMs) and finds PWMs maximizing the joint likelihood of pairwise occurrences. Since a limited number of Gibbs Sampling steps is performed to select motifs, many patterns are disregarded. Convergence to an optimal PWM is also not guaranteed.

MITRA-dyad [3] reduces the problem of searching for pairs of motifs to the one of finding a simple motif. It combines SMILE's tree-like structure and search, and the Winnower's graph construction for pruning the search space. MITRA relies on a mismatch tree. Although it will typically examine a smaller search space, analyzing a graph per node increases traversal time. RISO [2] is also based on SMILE and introduces box-links to store the information needed to jump from each motif to another and skip the corresponding nodes in the tree. A factor tree is used. Combined, these enhancements produce a significant reduction in both time and space.

ExMotif [10] extracts all frequent occurrences of a structured motif model with a given quorum. Sequences are represented by sorted lists of symbol occurrence positions. The positional join operation efficiently builds a list of occurrences when concatenating two sequences. Occurrences of simple/structured motifs are obtained by positional joins on the list of symbols/list of occurrences of simple motifs.

MUSA [6] generates all motif models of a given length. A matrix of co-occurrences is then constructed to depict the ε tolerant score of the most common configuration in the sequences. Biclustering is then applied to group models and identify structured motifs. Identification of all interesting correlations is not guaranteed but weak motifs, only revealed by exhaustive enumeration, can be found.

3   Methods

e-BiMotif is based on BlockMSA [9], an algorithm combining global sequence alignment and biclustering to produce local multiple sequence alignment (MSA). Note that, although local MSA and structured motif finding are two different problems, both aim at identifying locally conserved regions in sequences.

Given the definitions below, the problem of identifying structured motifs translates into that of finding optimal ordered sets of non-overlapping blocks composed of fragments occurring approximately in subsets of sequences. *e*-BiMotif addresses this problem in two steps: (**Step 1**) Identify candidate blocks (local conserved regions within subsets of sequences), performed as in [9]; (**Step 2**) Group subsets of sequences exhibiting a similar structure of candidate blocks and report the corresponding structured motifs, using *e*-BiMotif biclustering.

[**Fragment**]. For any sequence S_k (with $|S_k|$ symbols), $S_k[i..j]$ ($1 \leq i < j \leq |S_k|$) is a fragment, i.e., a contiguous subsequence, of sequence S_k starting at position i and ending at position j.

[**k-block**]. A k-block $b = \{f_1, f_2, ..., f_k\}$ is an ungapped alignment region conserved in a set of sequences S with k sequences with k equal length fragments, one from each k sequence in S.

[**Block similarity score**]. Given a k-block b, its similarity score is the sum of scores of all the $\binom{k}{2}$ pairwise combinations of fragments f_i, f_j ($1 \leq i < j \leq k$) from b: $Score_{block}(b) = \sum_{1 \leq i < j \leq k} Score(f_i, f_j)$, where $Score(f_i, f_j)$ is the score between f_i and f_j.

[**Chain**]. A set of k-blocks on a set of k sequences, $B = \{b_1, b_2, ...b_n\}$, where each b_i ($1 \leq i \leq n$) is a k-block, is called a chain if, for all pairwise combinations of blocks, (b_i, b_j) ($1 \leq i < j \leq n$), b_i and b_j are non-overlapping blocks.

[**Chain similarity score**]. The similarity score of a chain of k-blocks, $C = \{b_1, b_2, ..., b_n\}$, where each b_i ($1 \leq i \leq n$) is a k-block, is defined as the sum of the scores of all its k-blocks minus the gap penalties between them: $Score_{chain}(C) = \left[\sum_{1 \leq i < n} Score_{block}(b_i) - Gap(b_i, b_{i+1})\right] + Score_{block}(b_n)$

[**Structured motif**]. Given a set of chains on a given set of k sequences, $Chains = \{C_1, C_2, ..., C_n\}$, where each C_i ($1 \leq i \leq n$) is a chain, a structured motif is the chain C_i with the highest score.

3.1   Finding and Reporting Structured Motifs Using e-BiMotif

e-BiMotif differs from BlockMSA [9] in the biclustering step. The differences lie in two key steps explained in the subsections below: **(i)** mapping of the candidate block identification results to the biclustering formulation and **(ii)** biclustering method.

BlockMSA defines a binary matrix, where rows and columns represent candidate blocks and sequences, respectively (Fig. 1). Each element is either 1 or 0 depending on whether the block is in the sequence or not. BiMax [7] is then applied to the matrix. It uses a divide-and-conquer approach to extract biclusters composed of 1s not entirely contained in any other bicluster (inclusion-maximal biclusters).

Fig. 1 shows an example of MSA mapping to biclustering (left), the biclustering matrix obtained after applying BiMax (top right), the corresponding mapping from biclustering results back to MSA adapted from BlockMSA [9] (bottom-right). Note that in this example, BlockMSA would miss Block3 in bicluster V and sequence S7 in bicluster U. The original biclusters V and U could only be eventually recovered in the block assembly and post-processing step as explained below.

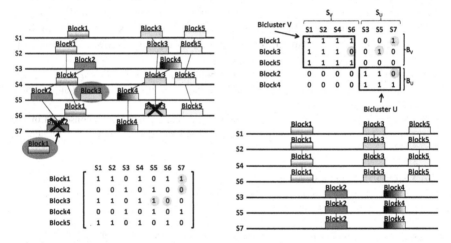

Fig. 1 Adapted figures from BlockMSA [9]. The original example was changed to present a case where the biclusters are not perfect (only composed of 1s) in order to show the advantages of *e*-BiMotif approach where errors (insertion, deletion, substitution) are considered in the biclustering method. The changes are highlighted: Block3 was inserted in S5 and deleted from S6; Block2 was replaced by Block1 in S7. These resulted in the substitutions 0 ↔ 1 highlighted in both matrices

3.1.1   Mapping Candidate Blocks to the Biclustering Formulation

We use the following MSA to biclustering mapping: each sequence is rewritten as the ordered set of the candidate blocks it contains, disregarding the non-conserved regions in-between. In this case, the alphabet of the new sequences holds the identifiers of all candidate blocks unraveled in the previous step (see Fig. 2).

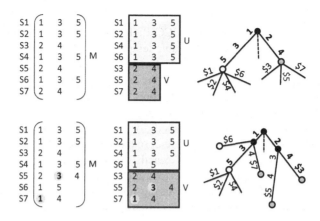

Fig. 2 Mapping for: the original example in BlockMSA [9] (top), and the adapted example in Fig. 1 (bottom). Also presented are the biclusters U and V in the generalized suffix tree used by *e*-BiMotif biclustering (details below). Nodes in light and dark grey identify the motifs

The problem of searching for structured motifs then translates into finding sequences with approximate candidate block structures, that is, identifying *e*-motifs and its biclusters (see definitions below) in the rewritten sequences. This formulation differs from the one in BlockMSA in three major points: **(i)** the non-conserved regions are disregarded; **(ii)** the order of blocks now plays an important role; **(iii)** we introduce an error threshold on the block structure. Essentially, we take advantage of block ordering granted by sequence alignment to restrict the problem to the one of finding contiguous subsequences with a maximum number of errors: *e*-motifs. Moreover, features **(i)** and **(iii)** allow us to eliminate the block assembly and post-processing step, used by BlockMSA to recover weakly conserved regions and merge blocks if they are within a relatively short distance. First, the inclusion of errors **(i)** enables to identify *e*-motif biclusters containing approximately conserved regions. Second, by disregarding non-conserved regions **(iii)** we guarantee that blocks can be included in the same *e*-motif bicluster regardless of the distance in-between.

[*e*-motif]. An *e*-motif is a sequence m of length $|m|$ occurring exactly or within an *e*-neighborhood in a set of sequences S at potentially different starting positions.

[*e*-neighborhood]. The *e*-neighborhood of a sequence S_k of length $|S_k|$, $N(e, S_k)$, is defined as the set of sequences S, such that: for each sequence S_i ($1 \leq i \leq |S|$) in S, $Hamming(S_k, S_i) \leq e$, where e is an integer such that $e \geq 0$. This means we need at most e operations (substitution, insertion, deletion) to obtain S_i from S_k and vice-versa.

3.1.2   Biclustering Method

e-BiMotif biclustering is based on e-CCC-Biclustering [4]. The major differences lie in: (i) the use of sequences of variable lengths, opposed to a matrix-like structure; (ii) the elimination of a matrix transformation, which was used to restrict the occurrences of a model to the same positions in the sequences.

e-BiMotif biclustering identifies and reports all maximal e-motif biclusters in a set of N sequences (see definitions below). In order to be valid, a given e-motif model m of length $|m|$ must occur exactly or approximately (any word in its e-neighborhood, $N(e,m)$) in at least q distinct sequences, where $2 \leq q \leq N$. Formally: given a set of N sequences S_i $(2 \leq N)$ and two integers $e \geq 0$ and $2 \leq q \leq N$, where e is the maximum number of errors allowed per e-motif occurrence and q the required quorum, e-BiMotif biclustering find all e-motif models m appearing in at least q distinct sequences of S_i. Biclusters U and V in the original (top) and adapted (bottom) examples in Fig. 2 are 0-motif biclusters and 1-motif biclusters, respectively.

[**e-motif bicluster**]. An e-motif bicluster is an e-motif m, a subset of sequences $s = \{s_1, s_2, ..., s_k\}$ from a set S $(|s| \leq |S|)$ and a set of initial and final positions $p = \{(i_1, f_1), (i_2, f_2), ..., (i_k, f_k)\}$, such that each pair of initial and final positions, (i_x, f_x) $(1 \leq x \leq k)$, identifies an exact or approximate (in the e-neighborhood) occurrence of e-motif m in sequence s_x.

[**Sequence-maximal e-motif bicluster**]. An e-motif bicluster is sequence-maximal if it cannot be added more sequences while maintaining the property referred in the definition of an e-motif.

[**Left/Right-maximal e-motif bicluster**]. An e-motif bicluster is left/right-maximal if its e-motif cannot be extended by adding a symbol to its beginning/end without losing sequences.

[**Maximal e-motif bicluster**]. An e-motif bicluster is maximal if it is sequence-left-right-maximal.

e-BiMotif biclustering starts by building a generalized suffix tree T for the set of sequences S_i. After further preprocessing, T is used to spell all valid e-motif models verifying two properties: (i) All prefixes of a valid model are also valid models; (ii) When $e = 0$, spelling a model leads to one node v in T such that $L(v) \geq q$. When $e > 0$, spelling a model leads to nodes $v_1, ..., v_k$ in T for which $\sum_{j=1}^{k} L(v) \geq q$. Since occurrences of a valid model are nodes in T, they are called node-occurrences. e-BiMotif identifies valid models by extending them in T and reporting their corresponding node-occurrences. Our definition of node-occurrence is an adaptation of the one used in e-CCC-Biclustering [4] to allow insertion and deletion errors:

[**Node-occurrence**]. A node-occurrence of a model m is a triple (v, v_{err}, p), where v is a node in T, v_{err} is the number of operations needed to transform m into the string-label of v and $p \geq 0$ is a position in T such that: (i) If $p = 0$ we are at node v; (ii) If $p > 0$ we are in $E(v)$, the edge from $father_v$ to v, in a point p between two symbols in $label(E(v))$ such that $1 \leq p \leq |label(E(v))|$.

e-BiMotif biclustering relies on three Lemmas supporting valid extensions of a model m to a new model m' by concatenating model m with a symbol α ($\alpha \in \Sigma$):

Lemma 1. $(v, v_{err}, 0)$ *is a node-occurrence of model* $m' = m\alpha$, *if and only if:*
(i) $(v, v_{err} - 1, 0)$ *is a node-occurrence of m* (**Deletion**).
(ii) $(father_v, v_{err}, 0)$ *is a node-occurrence of m and* $label(E(v)) = \alpha$ *or* $(father_v, v_{err},$ $|label(E(v))| - 1)$ *is a node-occurrence of m and* $label(E(v))[|label(E(v))|] = \alpha$ *(**Match**).
(iii) $(father_v, v_{err} - 1, 0)$ *is a node-occurrence of m and* $label(E(v)) = \beta \neq \alpha$ *or* $(v, v_{err} - 1, |label(E(v))| - 1)$ *is a node-occurrence of m and* $label(E(v))[|label(E(v))|]) = \beta \neq \alpha$
*(**Substitution**).*
(iv) $(father_v, v_{err} - 1, 0)$ *is a node-occurrence of m and* $|label(E(v))| = 1$ *or* $(v, v_{err} - 1, |label(E(v))| - 1)$ *is a node-occurrence of m and* $|label(E(v))| > 1$ *(**Insertion**).*

Lemma 2. $(v, v_{err}, 1)$ *is a node-occurrence of a model* $m' = m\alpha$, *if and only if:*
(i) $(v, v_{err} - 1, 1)$ *is a node-occurrence of m* (**Deletion**).
(ii) $(father_v, v_{err}, 0)$ *is a node-occurrence of m and* $label(E(v))[1] = \alpha$ *(**Match**).
(iii) $(father_v, v_{err} - 1, 0)$ *is a node-occurrence of m and* $label(E(v))[1] = \beta \neq \alpha$ *(**Substitution**).*
(iv) $(father_v, v_{err} - 1, 0)$ *is a node-occurrence of m and* $|label(E(v))| > 1$ *(**Insertion**).*

Lemma 3. (v, v_{err}, p), *where* $2 \leq p < |label(E(v))|$, *is a node-occurrence of a model* $m' = m\alpha$, *if and only if:*
(i) $(v, v_{err} - 1, p)$ *is a node-occurrence of m* (**Deletion**).
(ii) $(v, v_{err}, p - 1)$ *is a node-occurrence of m and* $label(E(v))[p] = \alpha$ *(**Match**).
(iii) $(v, v_{err} - 1, p - 1)$ *is a node-occurrence of m and* $label(E(v))[p] = \beta \neq \alpha$ *(**Substitution**).*
(iv) $(v, v_{err} - 1, p - 1)$ *is a node-occurrence of m* (**Insertion**).

Algorithm 1. *e*-BiMotif biclustering

    Input : $\{S_1, ..., S_N\}$, T, Σ, e, q_s, q_m
    Output: Maximal *e*-motif biclusters.
1 $modelsOcc \leftarrow \{\}$
2 computeRightMaximalBiclusters(Σ, e, q_s, q_m, $\{S_1, ..., S_N\}$, $modelsOcc$)
3 deleteNonLeftMaximalBiclusters($modelsOcc$)
4 **if** $e > 0$ **then**
5   deleteRepeatedBiclusters($modelsOcc$)
6 reportMaximalBiclusters($modelsOcc$)

`e-BiMotif` biclustering identifies and reports all maximal *e*-motif biclusters with at least q_s sequences and *e*-motif length of at least q_m in three steps: **(i)** Compute all models corresponding to sequence-right-maximal *e*-motif biclusters using Proc. `computeRightMaximalBiclusters` [4] and an adaptation of Proc. `spellModels` in [4]; **(ii)** Delete models identifying non left-maximal *e*-motif biclusters (Proc. `deleteNonLeftMaximalBiclusters` [4]); **(iii)** Delete models representing same *e*-motif biclusters (Proc. `deleteRepeatedBiclusters` [4]). Proc. `spellModels` traverses the tree to spell valid models and their node-occurrences (see pseudocode below). Proc. `extendModelFromNode` and Proc. `extendModelFromBranch` (Appendix A) perform extension of valid models.

Procedure spellModels

Input: Σ, e, q_s, q_m, $modelsOcc$, T_{right}, m, $length_m$, Occ_m, Ext_m, $father_m$,
       $numberOfSeqOcc_{father_m}$

1 keepModel(q_s, q_m, $modelsOcc$, T_{right}, m, $length_m$, Occ_m, $father_m$, $numberOfSeqOcc_{father_m}$)
   if $length_m < maximumModelLength$ **then**

2   **foreach** *symbol* α *in* Ext_m **do**

3          $maxSeq \leftarrow 0$

4          $minSeq \leftarrow \infty$

5   **if** α *is not a string terminator* **then**

6              $Colors_{m\alpha} \leftarrow \{\}$

7   **if** $e > 0$ **then**

8   **for** i *from* 1 *to* N **do**

9                     $Colors_{m\alpha}[i] \leftarrow 0$

10          $Ext_{m\alpha} \leftarrow \{\}$

11          $Occ_{m\alpha} \leftarrow \{\}$

12   **foreach** *node-occ* (v, v_{err}, p) *in* Occ_m **do**

13   removeNodeOccurrence((v, v_{err}, p), Occ_m)

14   **if** $p = 0$ **then**

15   extendModelFromNode(T_{right}, (v, v_{err}, p), α, Occ_m, $Occ_{m\alpha}$,
                    $Colors_{m\alpha}$, $Ext_{m\alpha}$, $maxSeq$, $minSeq$, 0)

16   **else**

17   extendModelFromBranch(T_{right}, (v, v_{err}, p), α, Occ_m, $Occ_{m\alpha}$,
                    $Colors_{m\alpha}$, $Ext_{m\alpha}$, $maxSeq$, $minSeq$, 0)

18   **if** *modelHasQuorum($maxSeq$, $minSeq$, $Colors_{m\alpha}$, q_s)* **then**

19   spellModels(Σ, e, q_s, q_m, $modelsOcc$, T_{right}, $m\alpha$, $length_m + 1$, $Occ_{m\alpha}$,
              $Ext_{m\alpha}$, m, $numberOfSeqOcc_m$)

3.2 Complexity Analysis of e-BiMotif

Step 1 (candidate block identification) takes $O(N^2 a^2) + O(Nfa)$ time, where N is the number of sequences, f is the number of seed fragments and a is the average sequence length in an unaligned region [9]. In Step 2 (biclustering), Proc. computeRightMaximalBiclusters is $O(NLB^e)$, Proc. deleteNonLeft MaximalBiclusters is $O(lLB^e)$, and Proc. deleteRepeatedBiclusters and Proc. report MaximalBiclusters are $O(NLB^e)$, where L is the sum of the lengths of the sequences, l is the length of the longest sequence, B is the size of the alphabet corresponding to the number of candidate blocks, and e is the maximum number of errors. Thus, Step 2 takes $O(NLB^e)$ when $N > l$ and $O(lLB^e)$ otherwise.

4 Conclusions and Future Work

We propose *e*-BiMotif to identify structured motifs in multiple sequences. This biclustering approach to motif finding efficiently unravels all approximately conserved blocks while straightforwardly disregarding non-conserved regions in-between.

Currently, *e*-BiMotif uses a candidate block identification step outlined in [9]. However, it remains unclear if global alignment is the best strategy to successfully identify locally conserved regions. In fact, this is the main motivation for the existence of local alignment methods. Moreover, the greedy clustering approach used for candidate block construction does not lead to an optimal solution. In this context, and since each *e*-motif actually represents a locally conserved region spanning a subset of sequences (an *e*-motif is a candidate block), it would be worth to replace sequence alignment by an *e*-BiMotif strategy using the original sequences.

The time complexity of *e*-BiMotif biclustering is $O(NL_o)$ and $O(4^eNL_o)$ to identify candidate blocks with perfectly and approximately conserved regions, respectively, outperforming [9] (N being the number of sequences, L_o the sum of lengths of the original sequences and e the number of errors). Moreover, *e*-BiMotif biclustering guarantees all locally conserved regions are found, while the combined approach of pairwise alignment and greedy clustering does not; there is no restrictions on the width of conserved regions, a major advantage when prior knowledge is not available; and the length of each conserved region is maximized, avoiding the generation of an excessive and unnecessary number of overlapping candidate blocks.

BiMax biclustering used in BlockMSA [9] takes $O(BNmin(B,N)\beta)$ to identify perfect biclusters, where B is the number of candidate blocks and β is the number of biclusters. *e*-BiMotif biclustering is $O(\max\{NL_rB^e, l_rL_rB^e\})$, where L_r is the total length of the rewritten sequences and l_r the length of the longest rewritten sequence. The biclustering approach in [9] is not able to consider errors or gaps between blocks, thus requiring an additional greedy block assembly and post-processing step.

Acknowledgements. This work was partially supported by project ARN, PTDC/EIA/67722/2006, funded by FCT. The work of JPG was supported by FCT grant SFRH/BD/36586/2007.

References

1. Buhler, J., Tompa, M.: Finding motifs using random projections. Journal of Computational Biology 9(2), 225–242 (2002)
2. Carvalho, A., Freitas, A., Oliveira, A., Sagot, M.F.: An efficient algorithm for the identification of structured motifs in DNA promoter sequences. IEEE/ACM Transactions on Computational Biology and Bioinformatics 3(2), 126–140 (2006)
3. Eskin, E., Pevzner, P.: Finding composite regulatory patterns in DNA sequences. Bioinformatics 18(Sup. I), S354–S363 (2002)
4. Madeira, S., Oliveira, A.: An efficient biclustering algorithm for finding genes with similar patterns in time-series expression data. Algorithms in Molecular Biology 4, 8 (2009)
5. Marsan, L., Sagot, M.F.: Extracting structured motifs using a suffix tree - algorithms and application to promoter consensus identification. Journal of Computational Biology 7, 345–354 (2000)
6. Mendes, N., Casimiro, A., Santos, P., Sá-Correia, I., Oliveira, A., Freitas, A.: MUSA: a parameter free algorithm for the identification of biologically significant motifs. Bioinformatics 22(24), 2996–3002 (2006)
7. Prelič, A., Bleuler, S., Zimmermann, P., Wille, A., Bühlmann, P., Gruissem, W., Hennig, L., Thiele, L., Zitzler, E.: A systematic comparison and evaluation of biclustering methods for gene expression data. Bioinformatics 22(9), 1122–1129 (2006)
8. Thakurta, D., Stormo, G.: Identifying target sites for cooperatively binding factors. Bioinformatics 17(7), 608–621 (2001)
9. Wang, S., Gutell, R., Miranker, P.: Biclustering as a method for RNA local multiple sequence alignment. Bioinformatics 23(24), 3289–3296 (2007)
10. Zhang, Y., Zaki, M.: EXMOTIF: efficient structured motif extraction. Algorithms for Molecular Biology 1(21), 126–140 (2006)

A. *e*-BiMotif: Algorithmic Details

Procedure extendModelFromNode

Input: T_{right}, (v, v_{err}, p), α, Occ_m, $Occ_{m\alpha}$, $Colors_{m\alpha}$, $Ext_{m\alpha}$, $maxSeq$, $minSeq$, $level$

1   **if** $level = 0$ **then**
2   **if** $v_{err} < e$ **then**
3   extendModel(T_{right}, α, Occ_m, $Occ_{m\alpha}$, $(v, v_{err} + 1, p)$, $Colors_{m\alpha}$, $Ext_{m\alpha}$, $maxSeq$, $minSeq$, $level$, false)
4   **if** v *is an internal node* **then**
5   **foreach** *child son of node* v **do**
6   **if** $label(E(son))[1]$ *is not a string terminator* **then**
7   **if** $|label(E(son))| > 1$ **then**
8           $p_{son} \leftarrow 1$
9   **else**
10          $p_{son} \leftarrow 0$
11   tryExtension(T_{right}, (v, v_{err}, p), son, p_{son}, β, α, Occ_m, $Occ_{m\alpha}$, $Colors_{m\alpha}$, $Ext_{m\alpha}$, $maxSeq$, $minSeq$, $level$)

Procedure extendModelFromBranch

Input: T_{right}, (v, v_{err}, p), α, Occ_m, $Occ_{m\alpha}$, $Colors_{m\alpha}$, $Ext_{m\alpha}$, $maxSeq$, $minSeq$, $level$

1   **if** $level = 0$ **then**
2   **if** $v_{err} < e$ **then**
3   extendModel(T_{right}, α, Occ_m, $Occ_{m\alpha}$, $(v, v_{err} + 1, p)$, $Colors_{m\alpha}$, $Ext_{m\alpha}$, $maxSeq$, $minSeq$, $level$, false)
4   **if** $label(E(v))[p + 1]$ *is not a string terminator* **then**
5   **if** $|label(E(v))| > p$ **then**
6       $p_{new} = p + 1$
7   **else**
8       $p_{new} = 0$
9   tryExtension(T_{right}, (v, v_{err}, p), son, p_{son}, β, α, Occ_m, $Occ_{m\alpha}$, $Colors_{m\alpha}$, $Ext_{m\alpha}$, $maxSeq$, $minSeq$, $level$)

Procedure findMinimumError

Input: $matchError$, $substitutionError$, $insertionError$, $deletionError$, β, α

1   $minErr \leftarrow insertionError$
2   $minErr \leftarrow \min(minErr, deletionError)$
3   **if** $\beta = \alpha$ **then**
4     $minErr \leftarrow \min(minErr, matchError)$
5   **else**
6     $minErr \leftarrow \min(minErr, substitutionError)$
7   **return** $minErr$

Procedure tryExtension

Input: T_{right}, (v, v_{err}, p), son, p_{son}, β, α, Occ_m, $Occ_{m\alpha}$, $Colors_{m\alpha}$, $Ext_{m\alpha}$, $maxSeq$, $minSeq$, $level$

1 **if** Occ_m *is not empty* **then**
2     $(x, x_{err}, p_x) \leftarrow getNextNodeOccurrence(Occ_m)$
3   **if** $x = son$ *and* $p_x = p_{son}$ **then**
4   removeNodeOccurrence$((x, x_{err}, p_x), Occ_m)$
5        $son_{err} \leftarrow$ findMinimumError$(v_{err}, v_{err} + 1, v_{err} + 1, x_{err} + 1, \beta, \alpha)$
6   **if** $son_{err} \leq e$ **then**
7   extendModel$(T_{right}, \alpha, Occ_m, Occ_{m\alpha}, (son, son_{err}, p_{son}), Colors_{m\alpha}, Ext_{m\alpha}, maxSeq, minSeq, level, true)$
8   **else**
9   removeChildrenNodeOccs(Occ_m, son, p_{son})
10   **return**
11 $son_{err} \leftarrow$ findMinimumError$(v_{err}, v_{err} + 1, v_{err} + 1, \beta, \alpha)$
12 extendModel$(T_{right}, \alpha, Occ_m, Occ_{m\alpha}, (son, son_{err}, p_{son}), Colors_{m\alpha}, Ext_{m\alpha}, maxSeq, minSeq, level, false)$

Procedure extendModel

Input: T_{right}, α, Occ_m, $Occ_{m\alpha}$, $Colors_{m\alpha}$, (n, n_{err}, p), $Ext_{m\alpha}$, $maxSeq$, $minSeq$, $level$ $doRecursion$

1 addNodeOccurrence$((n, n_{err}, p), Colors_{m\alpha})$
2 updateMaxSeqMinSeq$(maxSeq, minSeq, n)$
3 **if** $e > 0$ **then**
4   updateColors$(Colors_{m\alpha}, n)$
5 **if** $n_{err} = e$ **then**
6   **if** $p > 0$ **then**
7   **if** $label(E(n))[p + 1]$ *is not a string terminator* **then**
8   addSymbol$(label(E(n))[p + 1], Ext_{m\alpha})$
9   **if** $doRecursion = true$ **then**
10   extendModelFromBranch$(T_{right}, (n, n_{err}, p), \alpha, Occ_m, Occ_{m\alpha}, Colors_{m\alpha}, Ext_{m\alpha}, maxSeq, minSeq, level + 1)$
11   **else**
12   **if** n *is an internal node* **then**
13   **foreach** *child son of node n* **do**
14   **if** $label(E(n))[1]$ *is not a string terminator* **then**
15   addSymbol$(label(E(n))[1], Ext_{m\alpha})$
16   **if** $doRecursion = true$ **then**
17   extendModelFromNode$(T_{right}, (n, n_{err}, p), \alpha, Occ_m, Occ_{m\alpha}, Colors_{m\alpha}, Ext_{m\alpha}, maxSeq, minSeq, level + 1)$
18 **else**
19   addAllSymbols$(\Sigma, Ext_{m\alpha})$
20   **if** $doRecursion = true$ **then**
21   **if** $p > 0$ **then**
22   extendModelFromBranch$(T_{right}, (n, n_{err}, p), \alpha, Occ_m, Occ_{m\alpha}, Colors_{m\alpha}, Ext_{m\alpha}, maxSeq, minSeq, level + 1)$
23   **else**
24   extendModelFromNode$(T_{right}, (n, n_{err}, p), \alpha, Occ_m, Occ_{m\alpha}, Colors_{m\alpha}, Ext_{m\alpha}, maxSeq, minSeq, level + 1)$

Applying a Metabolic Footprinting Approach to Characterize the Impact of the Recombinant Protein Production in *Escherichia coli*

Sónia Carneiro, Silas G. Villas-Bôas, Isabel Rocha, and Eugénio C. Ferreira

Abstract. In this study metabolic footprinting was applied to evaluate the metabolic consequences of protein overproduction at slow growth conditions ($\mu = 0.1$ h^{-1}). The extracellular metabolites detected by gas chromatography-mass spectrometry characterized the metabolic footprints before and after the induction of the recombinant protein production (i.e. pre- and post-induction phases). Metabolic footprinting enabled the discrimination between the two growth phases and exposed significant alterations in the extracellular milieu during the recombinant process.

1 Introduction

Escherichia coli has been exploited for the production of a variety of products, especially recombinant proteins with pharmaceutical applications. Vast efforts have been made to improve the productivity of such bioprocesses, like the optimization of operational conditions, medium composition and the implementation of monitoring and control strategies [11,12,21]. However, the overproduction of these recombinant products often causes cellular stress events that result in slow growth and eventually cessation of growth [2,3,8,18]. The rapid exhaustion of essential metabolic precursors and cellular energy due to the expression of recombinant proteins may result in the imbalance of the metabolic load in the host cell, also called metabolic burden [9]. It is believed that the withdrawal of the intermediates that serve as biochemical precursors explains the decreasing tricarboxylic acid (TCA) cycle activity and consequent acetate production that has been reported in many works [1,5-7,10,22,23,25].

In recent years, various high-throughput experimental techniques have been used to understand the physiological behavior of cells during the operation of these bioprocesses and to develop strategies to overcome some of these limitations, like the design of improved strains that maximize the yield or productivity

Sónia Carneiro · Isabel Rocha · Eugénio C. Ferreira
IBB - Institute for Biotechnology and Bioengineering, Centre of Biological Engineering, University of Minho, Campus de Gualtar, 4710-057 Braga, Portugal

Silas G. Villas-Bôas
School of Biological Sciences, The University of Auckland, 3A Symonds Street, Auckland 1142, New Zealand

M.P. Rocha et al. (Eds.): IWPACBB 2010, AISC 74, pp. 193–200, 2010.
springerlink.com   © Springer-Verlag Berlin Heidelberg 2010

of recombinant proteins [4,14]. Transcriptome and proteome analyses have been widely used to investigate the stress response mechanisms associated with over-production of recombinant proteins in *E. coli*, but so far metabolomic approaches were scarcely exploited in the characterization of recombinant cultures. Metabolic footprinting, i.e. the analysis of the entire set of metabolites released from cells into the extracellular medium, can be an effective method to characterize the metabolic state of cells at diverse environmental conditions. The secretion of me-tabolites during the production of recombinant proteins may reflect the adjustment of the intracellular metabolism in response to the imposed metabolic demands [13], since the intracellular accumulation of certain metabolites due to metabolic imbalances will most probably result in their excretion. Therefore, metabolic foot-printing can represent an invaluable tool to generate key information that, together with other experimental data, will help in the optimization of recombinant cultures.

In this study, we investigated the usefulness of metabolic footprinting to assess the impact of the induction of recombinant protein production in *E. coli* cells. To avoid overlapping cellular responses that could be triggered, for example by meta-bolic overflow, cellular growth was maintained at low rates through the control of the glucose feeding profile.

2 Material and Methods

2.1 Growth Conditions

The *E. coli* strain W3110 (F-, *LAM*-, *IN*[*rrn*D-*rrn*E]1, *rph*-1) was transformed with the cloned pTRC-HisA-AcGFP1 plasmid encoding the production of the re-combinant AcGFP1 protein. The *gfp* gene was amplified from the pAcGFP1 plasmid (from Clontech) that encodes for the green fluorescent protein AcGFP1, a derivative of AcGFP from *Aequorea coerulescens*.

Cells were first grown in a shake flask pre-culture using minimal medium con-sisting of 5 g·kg^{-1} of glucose, 6 g·kg^{-1} of Na$_2$HPO$_4$, 3 g·kg^{-1} of KH$_2$PO$_4$, 0.5 g·kg^{-1} of NaCl, 1 g·kg^{-1} of NH$_4$Cl, 0.015 g·kg^{-1} of CaCl$_2$, 0.12 g·kg^{-1} of MgSO$_4$.7H$_2$O, 0.34 g·kg^{-1} of thiamine, 2 mL·kg^{-1} of trace-element solution (described elsewhere [16], 2 mL·kg^{-1} of vitamins solution (described elsewhere [16]), 20 mg·kg^{-1}of L-isoleucine and 100 mg·kg^{-1} of ampicillin. For fed-batch cultures, cells were there-after transferred to a fermenter with the same minimal medium, except glucose. The feeding medium consisted of 50 g·kg^{-1} of glucose, 10 g·kg^{-1} of NH$_4$Cl, 4 g·kg^{-1} of MgSO$_4$.7H$_2$O, 20 mg·kg^{-1}of L-isoleucine and 100 mg·kg^{-1} of ampicillin.

Fed-batch fermentation was conducted in a 5L fermenter (Biostat MD) with a working volume of 2 L at 37°C, pH 7 and dissolved oxygen (DO) above 30%. The induction of AcGFP1 protein production was performed with 1.5 mM IPTG (iso-propyl β-D-thiogalactoside) when the culture reached an optical density (OD600nm) of 2.3. Fermentation conditions were monitored and controlled via a computer supervisory system. A closed-loop feeding control algorithm was em-ployed to maintain the growth rate (μ) constant in the fed-batch culture [17]. The algorithm is based on a Monod kinetic model using glucose as the only growth-limiting substrate and can be represented by the following equation:

$$F = \frac{X.\mu.W_R}{Y_{X/S}.S_f}$$

To maintain the specific growth rate (μ) at $0.1h^{-1}$, the feeding profile was computed based on the growth yield in glucose ($Y_{X/S}$) that was set to 0.35 and the concentration of glucose in the feed (S_f), kept at 50 $g\cdot kg^{-1}$. The culture medium weight (W_R) was measured online, while the biomass concentration (X) was estimated based on the initial concentration.

2.2 Analytical Methods

Samples were taken from the fermenter approximately every 30 minutes for the determination of OD600nm, AcGFP1 fluorescence, glucose and acetate concentrations and the GC-MS analysis of extracellular amino and nonamino organic acids. In order to determine the cell dry weight, 10 mL of broth were centrifuged at 10000 g for 20 min at 4°C, washed twice with deionized water and dried at 105 °C to constant weight. The production of AcGFP1 was determined by fluorescence measurements at a Jasco FP-6200 spectrofluorometer with excitation and emission wavelengths of 475 and 505 nm, respectively, a bandwidth of 10 nm and a high sensitivity response in 0.1 seconds. His-Tag purification of the AcGFP1 was performed with HiTrap columns (GE Healthcare Bio-Sciences AB) and the concentration was determined by the Bradford method using BSA as standard. The metabolic footprints were analyzed by GC-MS. After the lyophilization of 1 mL of each sample in triplicates, chemical derivatization was performed using the methyl chloroformate (MCF) method described elsewhere [26]. Samples were thereafter analyzed with a GC-MS system – a GC7890 coupled to an MSD5975 - (Agilent Technologies) equipped with a ZB-1701 GC capillary column, 30 m x 250 mm id x 0.15 mm (film thickness) with 5 m guard column (from Phenomenex), at a constant flow rate of 1.0 mL/min of helium. Samples (1 μL) were injected onto the column under pulsed splitless mode (1.8 bars until 1 min, 20 mL/min split flow after 1.01 min) and the detector was set with a scan interval of 1.47 seconds and m/z range of 38-650.

2.3 Data Processing and Statistical Analysis

The mass fragmentation spectrum was analyzed with the Automated Mass Spectral Deconvolution and Identification System (AMDIS) [24] to identify the metabolites matching the analytical chemical standards. The peak intensity values of the identified metabolites in the spectrum were normalized by the peak intensity of the internal standard (D-4-alanine) and the corresponding biomass concentration of the sample. Further data processing and statistical analysis were performed with

MultiExperiment Viewer (MeV) v4.5 [20]. The normalized peak intensity values were log$_2$ transformed and further computed using K-means clustering (KMC), hierarchical clustering (HCL), and principal component analysis (PCA). K-means method was used to group the metabolic profiles into k clusters, while hierarchical clustering distributed samples and metabolites into branched groups represented by a two dimensional tree. Euclidean distance metrics were used in both clustering methods. PCA was further used to visualize whether the samples could be differentiated based on their metabolic profiles.

3 Results

In this study the impact of the production of recombinant proteins in the *E. coli* metabolism was investigated. The resulting growth pattern (Figure 1) shows that the biomass formation was affected by the AcGFP1 production, since before the IPTG-induction the experimental growth rate (0.16 h^{-1}) was higher than the growth rate imposed by the feeding profile. In turn, after IPTG induction the estimated biomass concentration was closer to the experimentally determined, corresponding to a growth rate of approximately 0.09 h^{-1}. This suggests that the biomass yields from glucose ($Y_{X/S}$) in the two growth phases were considerably different. In the first phase (pre-induction) glucose entering the cell was used for growth and maintenance, while in the second phase (post-induction) glucose was also allocated to AcGFP1 formation.

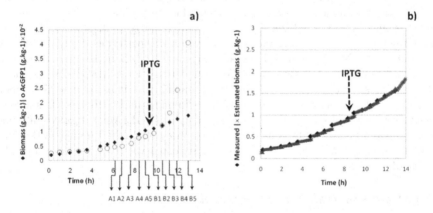

Fig. 1 Growth profile of the recombinant *E. coli* during the controlled fed-batch fermentation. a) The biomass formation and the AcGFP1 production were monitored in the pre- and post-induction phases. The dashed arrow indicates IPTG addition to the culture and the other arrows point to the samples analyzed in the GC-MS. Samples were identified by the letters A and B, corresponding to the pre- and post induction phases, respectively. b) The feeding profile generated by the model was periodically updated for the estimated biomass concentration

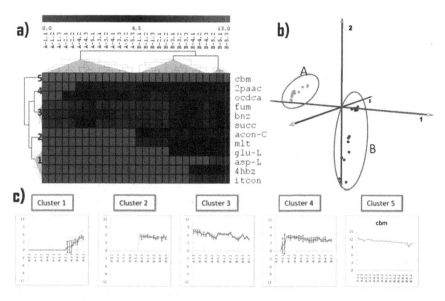

Fig. 2 Analysis of the metabolic footprints. a) Hierarchical clustering (HCL) defined clusters in the sample and metabolite dimensions. b) Principal component analysis (PCA) grouped samples from the pre-indution phase (group A) isolated from samples taken after IPTG-induction (group B). c) K-Means Clustering (KMC) gives the resulting clusters based on the KMC input parameters (50 interactions upper limit). Metabolite abbreviations: *cbm* - carbamic acid; *2paac* - 2-phenylaminoacetic acid; *ocdca* – stearic acid; *fum* – fumaric acid; *bnz* – benzoic acid; *succ* – succinic acid; *acon-C* – *cis*-aconitic acid; *mlt* – malonic acid; *glu-L* – L-glutamic acid; *asp-L* – L-aspartic acid; *4hbz* - hydroxybenzoic acid; *itcon* – itaconic acid

Since the experiments were performed at slow growth, no acetate was accumulated. However, unexpected metabolites were detected in the extracellular medium during the fed-batch experiment. According to the GC-MS results, important differences were identified in the metabolic footprints in both growth phases: pre-induction (phase A) and post-induction (phase B).

Results from HCL and PCA analyses (Figure 2a and 2b) show that samples taken during phases A and B presented distinct metabolic footprints. Although HCL shows that samples B5 cluster in a small sub-branch, it is clear that the metabolic footprints sampled before IPTG-induction (phase A) are distinct from those sampled after IPTG-induction (phase B). This was also confirmed by the PCA graph that assigned samples into two major groups corresponding to the growth phases. Further examination of the clusters generated by the K-means method (Fig. 2c) reveals that some metabolites presented similar profiles along the fermentation process. For example, cluster 2 was characterized by the extracellular accumulation of *cis*-aconitic acid (*acon-C*) and malonic acid (*mlt*) immediately after the IPTG induction, while metabolites within cluster 1 were secreted approximately two hours later, when the AcGFP1 production was most evident. Carbamic acid (*cbm*), involved in the production of a source of nitrogen for *E. coli* growth, was the only metabolite that did not cluster with any other metabolite.

4 Discussion and Conclusions

Both PCA and HCL analyses indicate that the recombinant protein production in *E. coli* has a high impact in the cellular metabolism. Several metabolites were secreted into the medium after IPTG induction, in particular metabolites that were collected within clusters 1 and 2. For the other metabolites, it is not possible to discern if their profiles were influenced by the production of recombinant protein or by the changing fed-batch conditions (biomass concentration increases during the experiment). However, metabolites collected in clusters 1 and 2 are more likely to be modulated by the recombinant process since the accumulation of these metabolites was only observed in phase B.

The accumulation of *cis*-aconitic acid, that participates in the TCA cycle and two other amino acids, which precursors are TCA intermediates, such as the L-glutamic and L-aspartic acids that are synthesized via the enzymatic conversion of the α-ketoglutaric and oxaloacetic acids, respectively, suggest that this pathway is notably affected by the production of the recombinant protein. As reported [15,19], the metabolic effects imposed by the plasmid maintenance and formation of the recombinant product were found to be associated with the metabolic burden caused by the withdrawal of some metabolic intermediates that serve as biochemical precursors. As a consequence, the activity of many biochemical pathways is affected and it is frequently observed a decrease in the TCA cycle activity and the production of by-products that, at high concentrations, might be toxic to the cells. Therefore, by unbalancing the activity of certain reactions, it is expected that some metabolites are accumulated and subsequently secreted into the extracellular medium. This supports the idea that the metabolic adjustments required to compensate the additional production of a recombinant product are not as efficient as desired from a bioprocess optimization perspective.

Although the metabolic footprint measurements are not entirely informative, since they do not provide a comprehensive analysis of the intracellular metabolic changes, they can be used as variables in a multivariate statistical process control, like dynamic principal component analysis, for the on-line bioprocess monitoring. Ultimately, this information can help to characterize the physiological state and culturing performance during recombinant processes. From an engineering point of view, it is crucial to operate these processes in order to achieve consistent and reproducible qualities by developing modeling strategies using real-time observation of process variables, and furthermore to detect in advance abnormal production conditions.

Acknowledgments. This work is partly funded by the Portuguese FCT (Fundação para a Ciência e Tecnologia) funded MIT-Portugal Program in Bioengineering (MIT-Pt/BS-BB/0082/2008). The work of Sónia Carneiro is supported by a PhD grant from FCT (ref. SFRH/BD/22863/2005).

References

1. Akesson, M., Hagander, P., Axelsson, J.P.: Avoiding acetate accumulation in *Escherichia coli* cultures using feedback control of glucose feeding. Biotechnol. Bioeng. 73, 223–230 (2001)
2. Bentley, W.E., Mirjalili, N., Andersen, D.C., et al.: Plasmid-encoded protein - the principal factor in the metabolic burden associated with recombinant bacteria. Biotechnol. Bioeng. 35, 668–681 (1990)
3. Bonomo, J., Gill, R.T.: Amino acid content of recombinant proteins influences the metabolic burden response. Biotechnol. Bioeng. 90, 116–126 (2005)
4. Bulter, T., Bernstein, J.R., Liao, J.C.: A perspective of metabolic engineering strategies: Moving up the systems hierarchy. Biotechnol. Bioeng. 84, 815–821 (2003)
5. Chou, C.P.: Engineering cell physiology to enhance recombinant protein production in *Escherichia coli*. Appl. Microbiol. Biotechnol. 76, 521–532 (2007)
6. Dittrich, C.R., Bennett, G.N., San, K.Y.: Characterization of the acetate-producing pathways in *Escherichia coli*. Biotechnol. Progr. 21, 1062–1067 (2005)
7. Eiteman, M.A., Altman, E.: Overcoming acetate in *Escherichia coli* recombinant protein fermentations. Trends Biotechnol. 24, 530–536 (2006)
8. Ganusov, V.V., Brilkov, A.V.: Estimating the instability parameters of plasmid-bearing cells. I. Chemostat culture. J Theor. Biol. 219, 193–205 (2002)
9. Glick, B.R.: Metabolic load and heterologous gene-expression. Biotechnol. Adv. 13, 247–261 (1995)
10. Jana, S., Deb, J.K.: Strategies for efficient production of heterologous proteins in *Escherichia coli*. Appl. Microbiol. Biotechnol. 67, 289–298 (2005)
11. Levisauskas, D., Galvanauskas, V., Henrich, S., et al.: Model-based optimization of viral capsid protein production in fed-batch culture of recombinant *Escherichia coli*. Bioprocess Biosyst. Eng. 25, 255–262 (2003)
12. Mahadevan, R., Doyle, F.J.: On-line optimization of recombinant product in a fed-batch bioreactor. Biotechnol. Progr. 19, 639–646 (2003)
13. Mapelli, V., Olsson, L., Nielsen, J.: Metabolic footprinting in microbiology: methods and applications in functional genomics and biotechnology. Trends Biotechnol. 26, 490–497 (2008)
14. Menzella, H.G., Ceccarelli, E.A., Gramajo, H.C.: Novel *Escherichia coli* strain allows efficient recombinant protein production using lactose as inducer. Biotechnol. Bioeng. 82, 809–817 (2003)
15. Ozkan, P., Sariyar, B., Utkur, F.O., et al.: Metabolic flux analysis of recombinant protein overproduction in *Escherichia coli*. Biochem. Eng. J. 22, 167–195 (2005)
16. Rocha, I., Ferreira, E.C.: On-line simultaneous monitoring of glucose and acetate with FIA during high cell density fermentation of recombinant *E. coli*. Anal. Chim. Acta. 462, 293–304 (2002)
17. Rocha, I., Veloso, A.C.A., Carneiro, S., et al.: Implementation of a specific rate controller in a fed-batch *E. coli* fermentation. In: Proceedings of the 17th World Congress The International Federation of Automatic Control (2008)
18. Rozkov, A., vignone-Rossa, C.A., Ertl, P.F., et al.: Characterization of the metabolic burden on *Escherichia coli* DH1 cells imposed by the presence of a plasmid containing a gene therapy sequence. Biotechnol. Bioeng. 88, 909–915 (2004)
19. Saeed, A.I., Sharov, V., White, J., et al.: TM4: a free, open-source system for microarray data management and analysis. Biotechniques 34, 374–378 (2003)

20. Saucedo, V.M., Karim, M.N.: Analysis and comparison of input-output models in a recombinant fed-batch fermentation. J. Ferment. Bioeng. 83, 70–78 (1997)
21. Shiloach, J., Kaufman, J., Guillard, A.S., et al.: Effect of glucose supply strategy on acetate accumulation, growth, and recombinant protein production by *Escherichia coli* BL21 ((λDE3) and *Escherichia coli* JM109. Biotechnol. Bioeng. 49, 421–428 (1996)
22. Shimizu, N., Fukuzono, S., Fujimori, K., et al.: Fed-batch cultures of recombinant *Escherichia coli* with inhibitory substance concentration monitoring. J. Ferment. Technol. 66, 187–191 (1988)
23. Stein, S.E.: An integrated method for spectrum extraction and compound identification from gas chromatography/mass spectrometry data. J. Am. Soc. Mass Spectrom. 10, 770–781 (1999)
24. Suarez, D.C., Kilikian, B.V.: Acetic acid accumulation in aerobic growth of recombinant *Escherichia coli*. Process Biochem. 35, 1051–1055 (2000)
25. Villas-Boas, S.G., Delicado, D.G., Akesson, M., et al.: Simultaneous analysis of amino and nonamino organic acids as methyl chloroformate derivatives using gas chromatography-mass spectrometry. Anal. Biochem. 322, 134–138 (2003)

Rbbt: A Framework for Fast Bioinformatics Development with Ruby

Miguel Vázquez, Rubén Nogales, Pedro Carmona, Alberto Pascual, and Juan Pavón

Abstract. In a fast evolving field like molecular biology, which produces great amounts of data at an ever increasing pace, it becomes fundamental the development of analysis applications that can keep up with that pace. The Rbbt development framework intends to support the development of complex functionality with strong data processing dependencies, as reusable components, and serving them through a simple and consistent API. This way, the framework promotes reuse and accessibility, and complements other solutions like classic APIs and function libraries or web services. The Rbbt framework currently provides a wide range of functionality from text mining to microarray meta-analysis.

1 Background

Molecular biology produces data in many different fields, and in great amounts. There is a great potential for developing new analysis methods and applications by either merging data from different sources, or by taking advantage of the large scale of several data repositories. These applications are, however, hindered by the complications in developing such systems, as they may require complex data processing pipelines just to set up the application environment. The fact is that many application that implement similar or closely related functionality end up having to roll up their own processing back-end pipelines, which are often very similar to one another and may actually account for a great portion of the development time. It would increase the turnout on these applications if these processing pipelines where implemented

Miguel Vázquez · Juan Pavón
Dep. Ingeniería del Software e Inteligencia Artificial, Universidad Complutense Madrid

Rubén Nogales · Pedro Carmona
Dep. Arquitectura de Computadores y Automática, Universidad Complutense Madrid

Alberto Pascual
Biocomputing Unit, National Center for Biotechnology-CSIC

M.P. Rocha et al. (Eds.): IWPACBB 2010, AISC 74, pp. 201–208, 2010.
springerlink.com   © Springer-Verlag Berlin Heidelberg 2010

in a clean and structured way such that they could be shared and reused between different applications.

In fact, this approach would not only allow for sharing just these pipelines, but would also allow for more sophisticated functionality to be exported as APIs, thanks to the possibility to easily install and process the supporting data. Things like identifier translation, gene mention recognition and normalization, or annotation enrichment analysis are good examples. The problem of providing APIs for these types of functionality, those that are very data dependant, is partially solved by allowing the necessary data files to be shipped along with the API. This solution, however, does not apply well when the data is to large, or is not static, either because it must be updated periodically, or because it must allow for some type of customization on how it is produced.

One solution to this problem is to build our own ad-hoc back-end systems, and export the functionality through the use of web services, such as those following the SOAP and REST communication protocols. This approach has several advantages, in particular, that the complexity of installing and administering the system falls only over the system developers and not over the API users. It does, however, have one important drawback with respect to classic APIs, other than possible reliability or latency problems, and that is that open source API can be modified and adapted to particular needs, while these web services are static, and their behaviour is entirely controlled by the developers and maintainers of the application.

The solution proposed in this work, and implemented by the Rbbt framework, is to develop the back-end processes in such a way that they can be shipped with the API so that they can install, automatically, the complete execution environment on the users system. This way, the users will have local access to the functionality while still maintaining the possibility of altering the code to fit their needs. Furthermore, by encapsulating all the complexity of the system behind a friendly to use top level API interface, the user can include this functionality, otherwise very complex, easily into their own scripts and applications, thus expanding the range of possibilities in their own analysis workflows.

Another important benefit of this approach is that the users will have direct access to the data used to provide the functionality, which can, in turn, allow for developing other functionality over this same data. Actually, by having access to the actual processing pipelines, these themselves can be adapted to process the data into any other format that suits better to the needs of the user.

By following a set of design practices, the process of implementing new analysis applications may leave behind a collection of such high-level functionality that can then be reused in other applications. In this fashion, the SENT and MARQ applications [10, 11] ended up producing the functionality in Rbbt, which where then expanded by including the functionality in Genecodis [6]. These functionality can now be used to develop new applications very fast, as much of the heavy lifting has already been done.

2   Implementation

Since one of the main objectives was to offer a cohesive top level API that provided access to all the functionality, we chose the Ruby language to implement that API. Ruby is a versatile language that can be used for small processing scripts, very much like Perl, but can also be used for web development using state-of-the-art frameworks such as the popular Ruby-on-Rails. This way, the usefulness of the API is maximized, since it can be used over the complete application stack, from small maintenance scripts to large web sites.

Ruby can also interface fairly easily with other tools such as the R environment or Java APIs. This allows the Rbbt API to wrap around these tools and offer their services through the main API, thus helping reduce the complexity of the user code by hiding it behind the API and data processing tools, which can take care of compiling and installing third party software.

Rbbt is designed to serve functionality that otherwise could not be included in classic API due to their strong data dependence. This means that the API must be able to find this data locally on the system, which in turn, requires a configuration tool that can take care of producing that data and installing it in the appropriate locations. We will refer to such locations as local data stores, and may include anything from data files, to third party software, or sandboxes with training and evaluation data to train models for machine learning methods. In fact, a few additional benefits arise by coupling the API functionality with these data stores, for instance, the possibility of implementing transparent caching strategies into the API.

The characteristics of data processing pipelines have similarities to compiling a software program, where certain files are produced by performing particular actions over their dependencies. So, for example, to produce a classifier for biomedical articles, one must build the model from the word vectors for positive and negative class articles, these word vectors are derived in turn from the text of these articles, which are themselves produced by downloading from PubMed the articles corresponding to the PubMed identifiers that are listed as positive and negative classes. Changing the way we compute the word vectors, for example, using a different statistic for feature selection, should only require updating the files downstream. The classic Make tool from UNIX offers a simple and clear way to define these dependencies and production steps. One drawback of Make is its rigid syntax; for this reason in Rbbt we use Rake, a Ruby clone of Make, that implements the same idea, but in 100% Ruby code, so that it can directly harness the full potential of the language, including, of course, the Rbbt libraries.

Data files in the local data stores are preferably saved in tab separated format, which is easy to parse, examine and modify using standard text editors. Only when performance, or some other factor, becomes an issue a database or a key value store should be used. This helps to maintain the interpretability and accessibility of the data.

3 Features

The Rbbt package includes a collection of **core functionality** to help accessing on-line data, and managing the local data stores. These include access to online repositories such as PubMed, Entrez, BioMart, The Gene Ontology, etc. Some of the resources accessed through the Rbbt API, such as PubMed articles, or gene records from Entrez, are cached to increase speed and reduce load on the remote servers. Access to general online resources which are likely to be updated periodically is also cached, but in a different type of non-persistent cache so that can be purged easily when performing system updates.

For managing the data stores, Rbbt includes functionality to work with tab-separated data files. These files often have an identifier in the first column and a list of properties as the successive columns. The Rbbt API includes functions to load this kind of files as a hash of arrays and merge several of such structures by matching arbitrary columns, which is used for example to build the identifier translation files from data in different sources such as biomart and bach download files from organisms specific databases.

Organism specific information. Rbbt uses a configuration file for each supported organism to list all the details for processing the basic information. One of such resources, for example, is the identifier file, which lists of the gene identifiers in some particular format, followed by other equivalent identifiers for that gene in other formats. In the case of yeast, for example, the native format is the SGD identifier, and other supported formats include Entrez Gene, Gene Symbol, or several Affymetrix probe id formats. These identifiers are retrieve from batch download files from organisms specific databases merged with informations gathered from BioMart and Entrez gene. Also, the Rbbt lists Gene Ontology associations for each gene, common synonyms for gene names as can appear on the literature, and PubMed articles associated to each gene derived from Entrez GeneRIF or GO association files. This files are the seed for the rest of Rbbt functionality for that organism, from simple identifier translation to automatic processing of GEO datasets, or models for gene mention and normalization, it all can be processed automatically once these files are in place, which is done by just complete these simple configuration files. Rbbt currently support 8 model organisms: *Homo sapiens, Mus musculus, Rattus norvegicus, Saccharomyces cerevisiae, Schizosaccharomyces pombe, Caenorhabditis elegans, Arabidopsis thaliana,* and *Candida albicans*. Other organisms can be easily included by just completing the information in their configurations files.

Text mining functionality in Rbbt includes building bag-of-words representations for documents, strategies for feature selection like TF-IDF or Kullback-Leibler divergence statistics, document classification, and several strategies for named entity recognition. In the last category are a general purpose named entity recognition system using regular expressions, which, coupled with the Polysearch thesaurus can be used to find mentions to diseases, cellular locations, and several other things. Rbbt implements a special purpose named entity recognition system for genes, which is also known as gene mention recognition [9], as well as a normalization [5] engine that can be used to map the mentions into the corresponding genes. Both the gene mention recognition and the normalization tasks have specific sandboxes inside the data stores with training and evaluation data to build and asses the models. Additionally, Rbbt wraps the Abner and Banner software APIs [7, 4] for gene mention, with the same top level interface, so that they can be used interchangeably with its in-house developed system. Our own gene mention system borrows most of its ideas from Abner and Banner; it is based on conditional random fields [3, 8], trained using CRF++ [2], but has a much simpler and flexible configuration system to define the features.

Microarray Analysis in Rbbt includes automatic processing pipelines that can compile a database of gene expression patterns from most of the thousands of datasets available in GEO [1], as well as other GEO series or any other microarray repository with a little configuration. This data consists of gigabytes of expression data that can be used in meta analysis applications. It also supports automatic translation of probe ids from different platforms into a organism based common reference format that can be used to perform analysis across experiments using different technologies. The API includes functionality to manage this automatic processing pipelines, an R function library in charge of the actual automatic processing, so that the data is easily used from this environment, and a collection of features to perform rank based comparison for content based retrieval of similar signatures.

4 Examples of Use

To illustrate the simplicity of using the Rbbt library we will use two example tasks, identifier translation of gene identifiers and gene mention recognition and normalization. For more complete examples check the applications sample directory, or the code for SENT and MARQ, which represent full blown applications using the complete Rbbt API.

In order to have access to this functionality the user must only retrieve the rbbt package, and run the processing pipelines to set up the necessary data and learning models, which is done using the configuration tool packaged in Rbbt.

The first example, in listing 1, is a command-line application that can translate gene identifiers for a given organism into Entrez Gene Ids. The input list of gene ids, which is provided through standard input one line each, may be specified in any of the supported gene id formats, which, for instance, in the case of *H. sapiens* includes more than 34 different formats such as Ensemble Gene Ids, RefSeq, or Affymetrix probes. The whole scripts boils down to loading the translation index for the organism and the selected target format (Entrez Gene Id in this case), and then using that index to resolve all our input ids.

The second example, in listing 2 (only relevant part is shown), is a command-line application that, given an organism, and a query string, performs the query in PubMed and finds mentions to genes in the abstracts of the matching articles. In this case the task boils down to loading the gene mention engine (*ner*), the normalization engine (*norm*), performing the query to find the list of articles (*pmid*), and then going through the articles performing the following two steps: use the gene mention

Listing 1 Translate Identifiers

```
require 'rbbt/sources/organism'
require 'rbbt/sources/pubmed'

usage =<<-EOT
  Usage: #{$0} organism

  organism = Sce, Rno, Mmu, etc. See 'rbbt_config organisms'

  You will need to have the organism installed. Example:
  'rbbt_config prepare organism -o Sce'.

  This scripts reads the identifiers from STDIN.

  Example:
  cat yeast_identifiers.txt | #{$0} Sce
EOT

organism = ARGV[0]

if organism.nil?
  puts usage
  exit
end

index =
  Organism.id_index(organism, :native => 'Entrez Gene Id')
STDIN.each_line{|l| puts "#{l.chomp} => #{index[l.chomp]}"}
```

engine to find potential gene mentions in the articles text, which is basically the title and abstract of the article, and then using the normalization engine to resolve the mentions into actual gene identifiers, using the article text to disambiguate between ties. The gene mention engine has several possible back-ends, Abner, Banner, and our own in-house development RNER, all based on conditional random fields. If RNER is used, the model for that specific organism must be trained, a process that, while been rather lengthy, is completely automated using the configuration tool, while still having room for customization, and needs only to be performed once. The normalization engine needs no training, and also has ample room for configuration.

5 Conclusions

Open source APIs offer an inestimable resource for software developers. However, due to the complexity of bioinformatics analysis, this sharing was unfeasible for many functionality due to the strong dependence in elaborate data processing pipelines required to set up the applications environment. The use of web services

Listing 2 Gene Mention Recognition and Normalization

```ruby
# Load data, this can take a few seconds
ner  = Organism.ner(organism, :rner) # Use RNER
norm = Organism.norm(organism)

# Query PubMed. Take only the last 'max' articles
pmids = PubMed.query(query, max.to_i)

# For each article:
PubMed.get_article(pmids).each{|pmid, article|

  # 1. Find mentions
  mentions = ner.extract(article.text)

  # 2. Normalize them
  codes = {}
  mentions.each{|mention|
    codes[mention] = norm.resolve(mention, article.text)
  }

  # 3. Print results
  puts pmid
  puts article.text
  puts "Mentions: "
  codes.each{|mention, list|
    puts "#{ mention } => #{list.join(", ")}"
  }
  puts
}
```

opens the possibility of offering an API over this functionality while hiding these complex processes from the API user. However, web servers may suffer from unreliability or latency, and have much less room for adapting them to different circumstances. Rbbt has shown that the approach of constructing the processing pipelines themselves so that they can set up the data stores automatically not only allows to provide these same API locally, but has a number of additional benefits, in particular, more options in terms of modifying and adapting the code, reuse of the data files, and reusing the processing pipelines for other tasks.

Rbbt has been successfully used, in whole and in part, in developing several production ready applications. These applications where developed very fast with a limited development group; Rbbt and agile development practices played a fundamental role in the their fast turnout.

The source code for the framework, as well as for several applications that use it, can be accessed at http://github.com/mikisvaz.

Acknowledgments

We acknowledge support from the project *Agent-based Modelling and Simulation of Complex Social Systems (SiCoSSys)*, which is financed by Spanish Council for Science and Innovation, with grant TIN2008-06464-C03-01. We also thank the support from the *Programa de Creacin y Consolidacin de Grupos de Investigacin UCM-BSCH, GR58/08*.

References

1. Barrett, T., Suzek, T.O., Troup, D.B., Wilhite, S.E., Ngau, W.C., Ledoux, P., Rudnev, D., Lash, A.E., Fujibuchi, W., Edgar, R.: NCBI GEO: mining millions of expression profiles–database and tools. Nucleic acids research 33(Database Issue), D562 (2005)
2. Kudo, T.: Crf++: Yet another crf toolkit (2005)
3. Lafferty, J.D., McCallum, A., Pereira, F.C.N.: Conditional Random Fields: Probabilistic Models for Segmenting and Labeling Sequence Data. In: Proc. of the Eighteenth International Conference on Machine Learning table of contents, pp. 282–289 (2001)
4. Leaman, R., Gonzalez, G.: Banner: An executable survey of advances in biomedical named entity recognition. In: Pacific Symposium on Biocomputing, vol. 13, pp. 652–663, Citeseer (2008)
5. Morgan, A., Lu, Z., Wang, X., Cohen, A., Fluck, J., Ruch, P., Divoli, A., Fundel, K., Leaman, R., Hakenberg, J., et al.: Overview of BioCreative II gene normalization. Genome Biology 9(Suppl. 2), S3 (2008)
6. Nogales-Cadenas, R., Carmona-Saez, P., Vazquez, M., Vicente, C., Yang, X., Tirado, F., Carazo, J.M., Pascual-Montano, A.: GeneCodis: interpreting gene lists through enrichment analysis and integration of diverse biological information. Nucleic Acids Research (2009)
7. Settles, B.: ABNER: an open source tool for automatically tagging genes, proteins and other entity names in text. Bioinformatics 21(14), 3191 (2005)
8. Settles, B., Collier, N., Ruch, P., Nazarenko, A.: Biomedical Named Entity Recognition using Conditional Random Fields and Rich Feature Sets. In: COLING 2004 International Joint workshop on Natural Language Processing in Biomedicine and its Applications (NLPBA/BioNLP) 2004, pp. 107–110 (2004)
9. Smith, L., Tanabe, L., Ando, R., Kuo, C.J., Chung, I.F., Hsu, C.N., Lin, Y.S., Klinger, R., Friedrich, C., Ganchev, K., et al.: Overview of BioCreative II gene mention recognition. Genome Biology 9(Suppl. 2), S2 (2008)
10. Vazquez, M., Carmona-Saez, P., Nogales-Cadenas, R., Chagoyen, M., Tirado, F., Carazo, J.M., Pascual-Montano, A.: SENT: semantic features in text. Nucleic Acids Research 37(Web Server issue), W153 (2009)
11. Vazquez, M., Nogales-Cadenas, R., Arroyo, J., Botas, P., Garca, R., Carazo, J.M., Tirado, F., Pascual-Montano, A., Carmona-Saez, P.: MARQ: an online tool to mine GEO for experiments with similar or opposite gene expression signatures (Under submission), http://marq.dacya.ucm.es

Analysis of the Effect of Reversibility Constraints on the Predictions of Genome-Scale Metabolic Models

José P. Faria, Miguel Rocha, Rick L. Stevens, and Christopher S. Henry

Abstract. Reversibility constraints are one aspect of genome-scale metabolic models that has received significant attention recently. This study explores the impact of complete removal of reversibility constraints on the gene essentiality and growth phenotype predictions generated using three published genome-scale metabolic models: the iJR904, the iAF1260, and the iBsu1103. In all three models, the accuracy in predicting essential genes declined significantly with the relaxation of reversibility constraints, while the accuracy in predicting nonessential genes increased only for the iJR904 and iAF1260 model. Additionally, the number of inactive reactions in all models declined substantially with the relaxation of the reversibility constraints. This study rapidly reveals the extent to which the reversibility constraints included in a metabolic model have been optimized, and it indicates those incorrect model predictions that may be repaired and those correct model predictions that may be broken by increasing the number of reversible reactions in a model.

1 Introduction

In recent years, Flux Balance Analysis (FBA) and genome-scale metabolic models are increasingly being used as a means of predicting the metabolic capabilities of an organism based on knowledge of the biochemical interactions taking place in the organism's metabolic pathways. These models are capable of predicting essential genes, growth phenotypes, culture conditions, and metabolic engineering strategies [1]. Additionally, the number of models available for analysis is rapidly growing. Currently, models have been published for over 20 microorganisms [2], with new high-throughput reconstruction methods emerging capable of producing thousands of draft models in a single year [3].

José P. Faria · Miguel Rocha
Dep. Informatics/ CCTC, University of Minho, Portugal

José P. Faria · Rick L. Stevens · Christopher S. Henry
Computation Institute, The University of Chicago, Chicago, IL, USA

José P. Faria · Rick L. Stevens · Christopher S. Henry
Argonne National Laboratory, Argonne, IL, USA

M.P. Rocha et al. (Eds.): IWPACBB 2010, AISC 74, pp. 209–215, 2010.
springerlink.com   © Springer-Verlag Berlin Heidelberg 2010

One aspect of genome-scale metabolic models that has received significant attention recently is the reversibility constraints governing the direction(s) of operation for all reactions included in the model. In the first genome-scale metabolic models, these constraints were based largely on data available in biochemical databases and knowledge of pathway directionality in well-known metabolic subsystems (e.g. glycolysis) [4, 5]. More recently, methods have emerged for predicting reaction reversibility/directionality based on thermodynamics and simple heuristic rules [6-8]. Finally, methods are available for adjusting reaction reversibility/directionality constraints to fit model predictions to available experimental phenotype data [9, 10]. All of this work demonstrates the impact that small targeted changes to model reversibility constraints have on the accuracy of model predictions. Here, we explore the impact of complete removal of reversibility constraints on the gene essentiality and growth phenotype predictions generated using three published genome-scale metabolic models: the *i*JR904 [4], the *i*AF1260 [11], and the *i*Bsu1103 [6]. The *i*JR904 and *i*AF1260 are both metabolic models of *E. coli* K12. The *i*JR904 model includes 931 reactions encompassing 904 ORFs; the *i*AF1260 model is a substantial expansion over the *i*JR904, including 2059 reactions and encompassing 1260 ORFs. The *i*Bsu1103 model was created for *B. subtilis* 168 and includes 1437 reactions encompassing 1103 ORFS. The *i*Bsu1103 model was optimized using the *GrowMatch* [9] method, in contrast to the *E. coli* models, which were manually optimized. This study is part of a larger study examining the impact of thermodynamic, regulatory, and reversibility constraints on the predictions from multiple genome-scale metabolic models.

2 Methods

2.1 Flux Balance Analysis (FBA)

Flux balance analysis (FBA) is a constraint-based simulation method used to define the limits on the metabolic capabilities of a microorganism [12-14]. In FBA, the interior of the cell is assumed to be in a quasi-steady-state, meaning that the net production/ consumption of each internal metabolite is zero. Based on this assumption, linear constraints are established on the flux through each reaction involved in the organism metabolism. Reaction fluxes are further constrained based on knowledge of the reversibility and directionality of the metabolic reactions, determined from thermodynamics [6-8]. A linear optimization is then performed with these constraints, such that a given metabolic objective function (often cell growth [15]) is maximized subject to the mass balance constraints, the reversibility constraints, and the availability of nutrients in the media. Gene knockouts may also be simulated by blocking all flux through metabolic reactions that are associated with the knocked out genes. Media conditions are set by restricting the compounds that can be consumed from the environment by the model reactions.

2.2 Classification of Reactions Using Flux Variability Analysis

Flux variability analysis (FVA) is an FBA-based method for characterizing the multiple feasible states of genome-scale metabolic models and for classifying the model reactions according to their behavior during simulated growth [16]. The reaction classification is derived from the minimization and maximization of flux through each model reaction while constraining the biomass production in the model to a minimal growth rate. Reactions with a minimum and maximum flux of zero are classified as *blocked* in the simulated conditions; reactions with a negative maximum flux or positive minimum flux are classified as *essential* in the simulated conditions; and all other reactions are classified as *active*.

3 Results

3.1 Impact of Reversibility Constrains on Model Accuracy

In order to study the effect of reversibility constraints on the accuracy of genome-scale metabolic model predictions, two genome-scale metabolic models of *E. coli* K12 (*i*JR904 [4] and *i*AF1260 [11]) and one genome-scale metabolic model of *B. subtilis* 168 (*i*Bsu1103 [6]) were utilized to predict the outcome of gene essentiality and Biolog growth phenotype experiments. These models were selected for analysis because they represent two of the most-well-studied prokaryotic organisms, one gram positive and one gram negative. Also, genome-wide gene essentiality and Biolog phenotyping array data are readily available for both of these organisms. Essentiality data is available for *E. coli* K12 in three distinct media conditions: Luria-Bertani media, glucose minimal media, and glycine minimal media [17, 18]. Essentiality data is also available for *B. subtilis* 168 in one culture condition: Luria-Bertani media [19].

The metabolic models were utilized to perform gene knockouts *in silico,* while simulating all culture conditions where experimental data is available. Knockouts were performed while enforcing and relaxing (by making all reactions reversible) the reversibility constraints included in each model. Predictions were then compared with experimental data to assess accuracy with and without reversibility constraints (Table 1). In all three models, the accuracy of gene essentiality predictions declined significantly with the relaxation of reversibility constraints, while the accuracy in predicting nonessential genes increased only for the *i*JR904 and *i*AF1260 models. This relaxation of reversibility constraints consists of making all model reactions reversible.

Biolog phenotyping arrays [20] have also been constructed and utilized to study the ability of *E. coli* K12 and *B. subtilis* 168 to metabolize 324 and 242 distinct carbon, nitrogen, phosphate, and sulfate sources respectively. The *i*JR904, *i*AF1260, and *i*Bsu1103 models were used to replicate these Biolog growth conditions *in silico,* while enforcing and relaxing the reaction reversibility constraints; all predictions were then compared against the experimental Biolog data (Table 1).

In these studies, the accuracy of all three models in predicting the metabolized Biolog nutrients improved with the relaxation of reversibility constraints, while accuracy in predicting un-metabolized nutrients declined. However, the improvement in the prediction of metabolized nutrients was much more substantial for the iJR904 and iAF1260 models than for the iBsu1103 model.

Table 1 Accuracy of Model Predictions with and without Reversibility Constraints

	iJR904[*]		iAF1260[*]		iBsu1103[*]	
Reversibility constraints	ON	OFF	ON	OFF	ON	OFF
Biolog conditions with growth	77/194 (40%)	99/194 (51%)	114/194 (59%)	130/194 (67%)	138/169 (82%)	142/169 (84%)
Biolog conditions with no growth	106/130 (82%)	79/130 (61%)	98/130 (75%)	71/130 (55%)	68/73 (93%)	50/73 (68%)
Essential metabolic genes	340/518 (66%)	229/518 (44%)	392/615 (64%)	99/615 (16%)	192/215 (89%)	166/215 (77%)
Non-essential metabolic genes	2000/2137 (94%)	2057/2137 (96%)	3053/3165 (95%)	3155/3165 (100%)	873/888 (98%)	873/888 (98%)
Overall accuracy	82.1%	80.2%	89.1%	84.2%	94.5%	91.5%

*Gene K.O. simulation results represent the aggregate of 3 media conditions (Luria-Bertani media, glucose minimal media, and glycine minimal media [17, 18]).

3.2 Impact of Reversibility Constraints on Reaction Behavior

Another measure of model quality is the number of inactive reactions in the model. Many reactions are supposed to be inactive during growth on certain conditions. For example, reactions involved in glycine metabolism should be inactive during growth on glucose minimal media. However, other reactions are inactive because they either exclusively lead to or are derived from a dead end in the metabolic network.

We utilized FVA to identify inactive reactions in the iJR904, iAF1260, and iBsu1103 models during minimal simulated growth in *complete media*. In *complete media*, any transportable nutrient is allowed to be taken up by the cell, making it the least restrictive media condition possible. The advantage of performing FVA on complete media is that this enables as many reactions as possible to be active since no uptake pathways are blocked. Thus, reactions identified as inactive in complete media represent those reactions that will never carry flux because they exclusively lead to or are derived from a dead-end metabolite. In some cases, these dead-ends can be eliminated with the relaxation of reversibility constraints. To identify these dead-ends, we repeated the FVA reaction classification to identify reactions that are no longer inactive with reversibility constraints relaxed (Figure 1). In all three models, the number of inactive reactions declined substantially with the relaxation of the reversibility constraints.

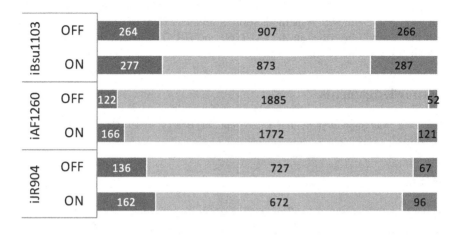

■ Inactive Reactions ▨ Active Reactions ■ Essential Reactions

Fig. 1 Number of inactive, active and essential reactions with reversibility constraints turned "ON" and "OFF"

4 Discussion

The results of our analysis of the effect of reversibility constraints on the accuracy of model predictions demonstrated that complete relaxation of reversibility constraints always results in a substantial decline in accuracy. However, the results also reveal that many cases where no growth is predicted and growth is observed (false negative predictions) can be corrected with the relaxation of reversibility constraints alone. More rigorous optimization techniques are available [9, 10] for identifying exactly which reactions should be made reversible to correct these predictions; however, this simple study provides a bulk estimate of how effective such efforts will be and it identifies the exact conditions on which such efforts should be applied. This study also reveals the correctly predicted zero-growth conditions that are vulnerable to being broken by the adjustment of reversibility constraints. Both pieces of information can be used to substantially simplify procedures for optimizing reaction reversibility constraints in models to fit experimental data.

Another interesting result can be derived from contrasting the effect of the reversibility constraints on the *i*JR904 and *i*AF1260 models versus the *i*Bsu1103 model. While in all three models, the number of false negative predictions declined with the relaxation of reversibility constraints, the decline was much more substantial in the *E. coli* models compared with the *i*Bsu1103 model. Meanwhile, the rise in false positive predictions with the relaxation of reversibility constraints was comparable in all three models. This ratio of errors corrected over errors created with the relaxation of reversibility constraints can be used as a measure of the extent to which a genome-scale metabolic model has been optimized. Thus, the reversibility rules in the *i*Bsu1103 model, which was optimized during reconstruction using the *GrowMatch* method, show a greater extent of optimality compared with the reversibility rules in the *i*AF1260 and *i*JR904 models, which underwent manual optimization only.

Acknowledgements

This work was supported by the U.S. Dept. of Energy under Contract DE- AC02-06CH11357.

References

1. Feist, A.M., Palsson, B.O.: The growing scope of applications of genome-scale metabolic reconstructions using Escherichia coli. Nat. Biotechnol. 26(6), 659–667 (2008)
2. Feist, A.M., Herrgard, M.J., Thiele, I., et al.: Reconstruction of Biochemical Networks in Microbial Organisms. Nature Reviews Microbiology 7(2), 129–143 (2009)
3. Henry, C.S., Xia, F., Stevens, R.L.: Application of High-Performance Computing to the Reconstruction, Analysis, and Optimization of Genome-scale Metabolic Models. Journal of Physics: Conference Series 180(012025) (2009)
4. Reed, J.L., Vo, T.D., Schilling, C.H., et al.: An expanded genome-scale model of Escherichia coli K-12 (iJR904 GSM/GPR). Genome Biology 4(9), 1–12 (2003)
5. Edwards, J.S., Palsson, B.O.: The Escherichia coli MG1655 in silico metabolic genotype: its definition, characteristics, and capabilities. Proceedings of the National Academy of Sciences of the United States of America 97(10), 5528–5533 (2000)
6. Henry, C.S., Zinner, J., Cohoon, M., et al.: iBsu1103: a new genome scale metabolic model of B. subtilis based on SEED annotations. Genome Biology 10, R69 (2009)
7. Kummel, A., Panke, S., Heinemann, M.: Systematic assignment of thermodynamic constraints in metabolic network models. BMC Bioinformatics 7, 512 (2006)
8. Henry, C.S., Jankowski, M.D., Broadbelt, L.J., et al.: Genome-scale thermodynamic analysis of Escherichia coli metabolism. Biophysical Journal 90(4), 1453–1461 (2006)
9. Kumar, V.S., Maranas, C.D.: GrowMatch: an automated method for reconciling in silico/in vivo growth predictions. PLoS Comput. Biol. 5(3) (2009) e1000308
10. Kumar, V.S., Dasika, M.S., Maranas, C.D.: Optimization based automated curation of metabolic reconstructions. BMC Bioinformatics 8, 212 (2007)
11. Feist, A.M., Henry, C.S., Reed, J.L., et al.: A genome-scale metabolic reconstruction for Escherichia coli K-12 MG1655 that accounts for 1260 ORFs and thermodynamic information. Mol. Syst. Biol. 3, 121 (2007)
12. Varma, A., Palsson, B.O.: Metabolic capabilities of Escherichia-coli.1. Synthesis of biosynthetic precursors and cofactors. Journal of Theoretical Biology 165(4), 477–502 (1993)
13. Varma, A., Palsson, B.O.: Metabolic capabilities of Escherichia-coli. 2. Optimal-growth patterns. Journal of Theoretical Biology 165(4), 503–522 (1993)
14. Papoutsakis, E.T., Meyer, C.L.: Equations and calculations of product yields and preferred pathways for butanediol and mixed-acid fermentations. Biotechnology and Bioengineering 27(1), 50–66 (1985)

15. Edwards, J.S., Ibarra, R.U., Palsson, B.O.: In silico predictions of Escherichia coli metabolic capabilities are consistent with experimental data. Nature Biotechnology 19(2), 125–130 (2001)
16. Chen, T., Xie, Z.W., Ouyang, Q.: Expanded flux variability analysis on metabolic network of Escherichia coli. Chinese Science Bulletin 54(15), 2610–2619 (2009)
17. Joyce, A.R., Reed, J.L., White, A., et al.: Experimental and computational assessment of conditionally essential genes in Escherichia coli. J Bacteriol 188(23), 8259–8271 (2006)
18. Baba, T., Ara, T., Hasegawa, M., et al.: Construction of Escherichia coli K-12 in-frame, single-gene knockout mutants: the Keio collection. Mol. Syst. Biol. 2, 2006 0008 (2006)
19. Kobayashi, K., Ehrlich, S.D., Albertini, A., et al.: Essential Bacillus subtilis genes. Proc. Natl. Acad. Sci. U. S. A 100(8), 4678–4683 (2003)
20. Bochner, B.R.: Global phenotypic characterization of bacteria. Fems Microbiology Reviews 33(1), 191–205 (2009)

Enhancing Elementary Flux Modes Analysis Using Filtering Techniques in an Integrated Environment

Paulo Maia, Marcellinus Pont, Jean-François Tomb, Isabel Rocha, and Miguel Rocha

Abstract. Elementary Flux Modes (EFMs) have been claimed as one of the most promising approaches for pathway analysis. These are a set of vectors that emerge from the stoichiometric matrix of a biochemical network through the use of convex analysis. The computation of all EFMs of a given network is an NP-hard problem and existing algorithms do not scale well. Moreover, the analysis of results is difficult given the thousands or millions of possible modes generated. In this work, we propose a new plugin, running on top of the OptFlux Metabolic Engineering workbench, whose aims are to ease the analysis of these results and explore synergies among EFM analysis, phenotype simulation and strain optimization.

1 Introduction

Over the last few years, a growing number of genome-scale metabolic models for different organisms have been reconstructed, based on the information contained in their annotated sequenced genomes, on the application of Bioinformatics tools and on physiological data. Given the major difficulties in obtaining the kinetic parameters for the whole set of reactions and also in reaching reliable regulatory information, most methods rely on the analysis of the network stoichiometry, using a constraint-based approach to reach some conclusions and to achieve the phenotypic simulation of the system.

Paulo Maia · Isabel Rocha
Dep. Biological Engineering / IBB-CEB, University of Minho, Braga, Portugal
e-mail: {paulo.maia, irocha}@deb.uminho.pt

Marcellinus Pont · Jean-François Tomb
E.I. DuPont De Nemours & Co., Inc, Dellaware, USA
e-mail: {marcellinus.pont, jean-francois.tomb}@usa.dupont.com

Miguel Rocha
Dep. Informatics/ CCTC, University of Minho, Braga, Portugal
e-mail: mrocha@di.uminho.pt

M.P. Rocha et al. (Eds.): IWPACBB 2010, AISC 74, pp. 217–224, 2010.
springerlink.com   © Springer-Verlag Berlin Heidelberg 2010

Metabolic models and stoichiometric network analysis have been successfully used to address tasks such as assessing the network's capabilities (e.g. determining maximal product yields), network design (e.g. studying the effects of adding/removing reactions), analysis of functional pathways, analysis of network consistency, flexibility or robustness, among others [13]. One of the major application fields of stoichiometric network analysis has been Metabolic Engineering (ME), a field that deals with designing organisms with enhanced capabilities regarding the productivities of desired compounds [12]. Recently, we have developed *OptFlux* [6], a new computational platform for ME, based on stoichiometric network analysis and constraint-based approaches. The platform implements the major methods related to phenotypical simulation using steady state approaches, such as Flux Balance Analysis (FBA) [3], where a flux distribution that obeys the constraints and maximizes a pre-defined biomass flux is obtained. Also, the platform provides algorithms for strain optimization (e.g. identifying the set of reactions to delete from a model in order to maximize a given objective function).

In contrast to this approach, and though also based on stoichiometric network analysis, the field of Pathway Analysis (PA) characterizes the complete space of admissible flux distributions. PA allows the analysis of meaningful routes involved in metabolic networks. In the recent past, two closely related approaches were developed; namely, Elementary Flux Modes (EFMs) [8] and Extreme Pathways (EPs) [7]. Both approaches aim to dissect metabolic networks into basic functional units and provide means to understand their behavior.

Given the importance of PA within the stoichiometric analysis framework, this paper reports on the development of a software tool that addresses the major tasks within PA. This tool is developed within the *OptFlux* project as a new plug-in. The main aims of this work are (i) to enhance the ME capabilities of the OptFlux platform using EFM analysis and (ii) to provide a graphical user interface to one of the most effective libraries in the computation of EFMs.

2   Elementary Modes

In this paper, we will use a purely structural analysis of biochemical networks, where the network is represented by: q internal compounds (metabolites), m reactions and a stoichiometric matrix N with dimensions $q \times m$. The rows of N correspond to the compounds, while the columns represent reactions; the elements of the matrix in row i and column j represent the stoichiometric coefficient of compound i in reaction j. The framework also allows the definition of external metabolites, thought to be sinks or sources, which lie outside the system. Also, reactions are defined to be reversible or irreversible. Figure 1 provides a simple example extracted from [13] with 6 internal metabolites and 10 reactions; also, this network has 4 external metabolites.

In a steady-state situation, the mass-balance in the network can be represented by the equation:

$$N\mathbf{v} = 0 \qquad (1)$$

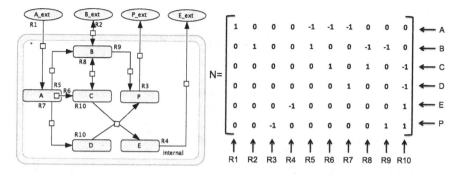

Fig. 1 The simple network used in this case study. $N : q \times m$ is the stoichiometric matrix where each row corresponds to one species and each column to one of the reactions. The reversibility information is given by $rev = \{R2, R8\}$ and $irrev = \{R1, R3, R4, R5, R6, R7, R9, R10\}$, where rev is the set of reversible reactions and $irrev$ is the set of irreversible ones. Also notice that $rev \cap irrev = \emptyset$

where $\mathbf{v}$ is a vector with the fluxes through the set of reactions. Irreversible reactions are represented by including constraints in the form:

$$v_i \geq 0 \tag{2}$$

The set of all flux vectors that obey the constraints imposed by equations 1 and 2 represent a convex polyhedral cone or a flux cone in stoichiometric studies.

In this context, Elementary Flux Modes (EFMs) are flux vectors composed of q elements $(e_1, e_2, \ldots, e_q)$, that fulfill the following conditions [8]:

1. **Pseudo steady state:** $N\mathbf{e} = 0$, i.e. all EFMs obey equation 1;
2. **Feasibility:** all irreversible reactions proceed in the forward direction, i.e. $e_i \geq 0$ if reaction i is irreversible (equation 2);
3. **Non-decomposability:** EFMs represent the minimal functional units in the network, therefore no reaction can be deleted from an EFM while maintaining a valid flux distribution (that satisfies equations 1 and 2)

This definition imposes the following important properties: (i) there is a unique set of EFMs for a given network; (ii) all feasible steady state flux distributions (that satisfy equations 1 and 2) are a nonnegative superposition of the set of EFMs in the network; (iii) when removing a reaction from the network, the set of EFMs for the new network is equal to the one for the original network but removing all EFMs that include the reaction that was deleted.

Extreme Pathways (EPs) are closely related to EFMs. They are a subset of the EFMs calculated over a reformulated network. Given the similarity of the two concepts, EPs will not be further addressed in this paper nor in the current version of the plug-in.

EFMs are a very useful tool for the analysis of a metabolic network, since they provide the portfolio of all elementary functional units. Thus, EFM analysis can be used, among others, in the following tasks: (i) identify all routes (pathways) that convert a substrate into a given product and also identify the ones with maximal yield; (ii) identify the importance of reactions in a given context and the correlated reactions; (iii) predict the effect of reaction deletions and viability of resultant mutants.

Since the number of EFMs grows exponentially with the network size, the computation of the whole set of EFMs for a metabolic network is a very hard problem. Several algorithms have been proposed to calculate EFMs (or EPs). They are mostly based on approaches that solve the equivalent problem of extreme ray enumeration from computational geometry. The developed algorithms that solve this problem are typically based on the double description method and can be grouped into two major approaches: the canonical basis approach [8] and the nullspace approach [16], which introduced simplifications in the algorithm and lead to an improved performance.

To further boost the performance, binary vectors were proposed [1] to store the processed reactions in each EFM, reducing memory demands and facilitating set operations. Also, rank computations were introduced to test elementarity in a divide and conquer strategy. Finally, Terzer and Stelling [14] introduced bit pattern trees to index subsets during elementarity testing and also the concept of candidate narrowing using a recursive enumeration approach.

When discussing software tools for EFM calculation, the METATOOL application is a reference [15]. METATOOL was developed in MatLab and has been quite popular since its first version in 1998. It is also included in other more comprehensive packages such as YANASquare [9] or CellNetAnalyzer [5]. More recently, the EFMTool was developed [14] using the Java language (but including a wrapper allowing it to run over MatLab). This is a library that implements a very efficient algorithm for EFM enumeration. This most effective approach is currently available through a command line interface. A major objective of our work is to provide a graphical user interface to enable easy access to the EFMTool.

3 Efm4Optflux – An EMA Plugin for the OptFlux Workbench

The need for a rational approach to analyze results generated by the EFM calculation algorithms was the driver behind the creation of a plug-in within the OptFlux ME platform. OptFlux is a freely available open-source software. It was developed in a modular fashion to facilitate the addition of new features (plug-in based architecture), it is compatible with the Systems Biology Markup Language (SBML) [2] and the layout information of Cell Designer [4]).

In the current version (2.0), the software accommodates several tools that have been developed for the analysis of metabolic models. It incorporates methods for phenotype simulation, such as Flux Balance Analysis, Minimization of Metabolic Adjustment [10] and Regulatory on/off minimization of metabolic flux changes [11], as well as strain optimization algorithms, such as Evolutionary Algorithms and Simulated Annealing. It also packages a suitable model visualization tool.

In this work, we present an extension to OptFlux, a plug-in that provides graphical access to the EFMTool calculation and analysis capabilities. The plugin delivers several interesting characteristics such as a simple graphical user interface (GUI) and state-of-the-art EFM calculation. It also allows filtering of the results based on an intuitive form, allowing the definition of patterns with presence/absence of external metabolites. The organization of the filtered results is done by grouping the EFMs with unique net conversions.

After the calculation of the EFMs, each EFM vector is multiplied by the submatrix component of the full stoichiometric matrix containing only the rows relative to the external metabolites, thus returning the vectors of net conversions. These are scanned and only unique conversions are maintained. All the results are kept in disk in order to ease the memory burden of some EFM computations. The EFMs are kept in one file, the net conversions in another and a lookup table is generated, in a third file, to correlate between them.

Furthermore, for each net conversion, the greatest common divisor is calculated to improve the reading of the conversion equation. To do so, all the coefficients have to be integers and therefore the EFM calculation is limited to using big integer arithmetics. In the filtering step, the software filters EFMs based on the presence/absence of external metabolites in the net conversions. Moreover, it can also order by yield, providing an input and output metabolite is provided for the computation. When the filters have been defined, the software scans the definition files for compatible conversions and their related EFMs.

The user can browse through the filtered conversions in an intuitive table that presents the conversion equation and yields and provides access to the related EFMs. The visualization of these EFMs is presented in a column-wise table, where each column corresponds to an EFM and each line to a reaction of the model.

Moreover, the user can export the EFM values to CellDesigner, if a valid CellDesigner SBML file had been previously loaded. The values for each reaction in the mode vector are represented by the thickness of the lines in the CellDesigner layout. This thickness varies between a user-defined range and it is relative to the value.

4   Case Study

We present the plugin capabilities, following a simple example from [13]. This network is presented in Figure 1. This example is provided in the project homepage (www.optflux.org) as a CellDesigner SBML file.

In order to compute the EFMs for this network, the following steps were required:

1. Load the file *Stelling_toy.xml* using the **New Project Wizard**. Select SBML, CellDesigner SBML and default options until the end of the wizard;
2. To execute the EFMs computation, access **Plugins** → **Elementary Modes** → **Compute Elementary Modes** and use all the default options. This computation is instantaneous in any regular PC;

3. Clicking the new datatype placed on the clipboard will launch a generic viewer. Press **Proceed to filtering** and a filtering dialog will appear. In this dialog we selected every filter as *Don't Care* and the yield as *A_ext → P_ext*.
4. A *FilteredEFMResults* instance will be placed on the clipboard. Clicking it will launch the viewer (Figure 2).

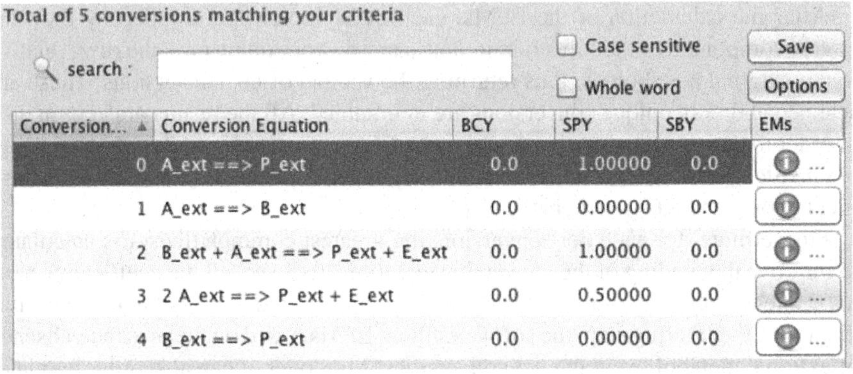

Conversion... ▲	Conversion Equation	BCY	SPY	SBY	EMs
0	A_ext ==> P_ext	0.0	1.00000	0.0	ⓘ ...
1	A_ext ==> B_ext	0.0	0.00000	0.0	ⓘ ...
2	B_ext + A_ext ==> P_ext + E_ext	0.0	1.00000	0.0	ⓘ ...
3	2 A_ext ==> P_ext + E_ext	0.0	0.50000	0.0	ⓘ ...
4	B_ext ==> P_ext	0.0	0.00000	0.0	ⓘ ...

Total of 5 conversions matching your criteria. search : ☐ Case sensitive ☐ Whole word — Save, Options

Fig. 2 Screenshot of the Filtered Results Viewer. A total of 5 unique conversions were discovered by the algorithm. These are listed together with the conversion equation and respective yields. BCY = Biomass/product Coupled Yield; SPY = Substrate to Product Yield; SBY = Substrate to Biomass Yield

The computation resulted in a total of 8 EFMs and 5 unique net conversions. The EFMs were exported to CellDesigner and the results were compared and validated with the ones obtained in [13]. A graphical representation of the layout in CellDesigner is presented in Figure 3.

5 Conclusion

Despite several incursions regarding the computation of EFMs, no real efforts have been made to ease analysis of the results. The proposed plug-in provides a seamless integration with the OptFlux ME workbench, thus providing a rational interface and smart filters to analyze the large number of EFMs that usually generated in such analysis approaches. There is, however, still much space for improvement, namely regarding the arithmetics allowed by the platform. The limitation to integers, though consistent with the display mode, is nevertheless, a limitation. Other ways of filtering and sorting the calculated EFMs are also currently under development.

Acknowledgements. The authors wish to thank the financial support of the Portuguese FCT for the Ph.D grant SFRH/BD/61465/2009 and the company Dupont under the scope of the *Dupont European University Support Program Award*.

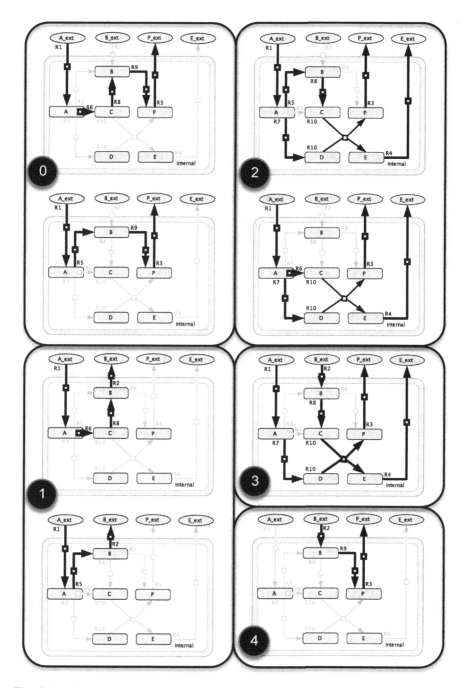

Fig. 3 Graphical representation of the EMs associated with each unique net conversion. The numbers are the Conversion IDs depicted in Figure 2. The conversions are the following: **0** :$A\_ext \rightarrow P\_ext$, **1** :$A\_ext \rightarrow B\_ext$, **2** :$2A\_ext \rightarrow P\_ext + E\_ext$, **3** :$B\_ext + A\_ext \rightarrow P\_ext + E\_ext$, **4** :$B\_ext \rightarrow P\_ext$

References

1. Gagneur, J., Klamt, S.: Computation of elementary modes: a unifying framework and the new binary approach. BMC Bioinformatics 5(175) (2004)
2. Hucka, M., Finney, A., Sauro, H.M., Bolouri, H., Doyle, J.C., Kitano, H., et al.: The systems biology markup language (SBML): a medium for representation and exchange of biochemical network models. Bioinformatics 19(4), 524 (2003)
3. Kauffman, K.J., Prakash, P., Edwards, J.S.: Advances in flux balance analysis. Curr. Opin. Biotechnol. 14, 491–496 (2003)
4. Kitano, H., Funahashi, A., Matsuoka, Y., Oda, K.: Using process diagrams for the graphical representation of biological networks. Nature biotechnology 23(8), 961–966 (2005)
5. Klamt, S., Saez-Rodriguez, J., Gilles, E.: Structural and functional analysis of cellular networks with cellnetanalyzer. BMC Systems Biology 1(2) (2007)
6. Rocha, I., Maia, P., Evangelista, P., Vilaça, P., Soares, S., Pinto, J.P., Nielsen, J., Patil, K.R., Ferreira, E.C., Rocha, M.: Optflux: an open-source software platform for in silico metabolic engineering. BMC Systems Biology (in Press, 2010)
7. Schilling, C.H., Letscher, D., Palsson, B.Ø.: Theory for the systemic definition of metabolic pathways and their use in interpreting metabolic function from a pathway-oriented perspective. Journal of Theoretical Biology 203(3), 229–248 (2000)
8. Schuster, S., Hilgetag, C.: On elementary flux modes in biochemical reaction systems at steady state. J. Biol. Syst. 2(2), 165–182 (1994)
9. Schwarz, R., Liang, C., Kaleta, C., Kuhnel, M., Hoffmann, E., Kuznetsov, S., Hecker, M., Griffiths, G., Schuster, S., Dandekar, T.: Integrated network reconstruction, visualization and analysis using yanasquare. BMC Bioinformatics 8(313) (2007)
10. Segre, D., Vitkup, D., Church, G.M.: Analysis of optimality in natural and perturbed metabolic networks. Proceedings of the National Academy of Sciences 99(23), 15112 (2002)
11. Shlomi, T., Berkman, O., Ruppin, E.: Regulatory on/off minimization of metabolic flux changes after genetic perturbations. Proceedings of the National Academy of Sciences of the United States of America 102(21), 7695 (2005)
12. Stephanopoulos, G., Aristidou, A.A., Nielsen, J.: Metabolic engineering principles and methodologies. Academic Press, San Diego (1998)
13. Szallasi, Z., Stelling, J., Periwal, V.: System modeling in cell biology: from concepts to nuts and bolts. MIT Press, Cambridge (2006)
14. Terzer, M., Stelling, J.: Large-scale computation of elementary flux modes with bit pattern trees. Bioinformatics 24(19), 2229–2235 (2008)
15. von Kamp, A., Schuster, S.: Metatool 5.0: fast and flexible elementary modes analysis. Bioinformatics 22(15), 1930–1931 (2006)
16. Wagner, C.: Nullspace approach to determine the elementary modes of chemical reaction systems. J. Phys. Chem. B 108, 2425–2431 (2004)

Genome Visualization in Space

Leandro S. Marcolino, Bráulio R.G.M. Couto, and Marcos A. dos Santos

Abstract. Phylogeny is an important field to understand evolution and the organization of life. However, most methods depend highly on manual study and analysis, making the construction of phylogeny error prone. Linear Algebra methods are known to be efficient to deal with the semantic relationships between a large number of elements in spaces of high dimensionality. Therefore, they can be useful to help the construction of phylogenetic trees. The ability to visualize the relationships between genomes is crucial in this process. In this paper, a linear algebra method, followed by optimization, is used to generate a visualization of a set of complete genomes. Using the proposed method we were able to visualize the relationships of 64 complete mitochondrial genomes, organized as six different groups, and of 31 complete mitochondrial genomes of mammals, organized as nine different groups. The prespecified groups could be seen clustered together in the visualization, and similar species were represented close together. Besides, there seems to be an evolutionary influence in the organization of the graph.

1 Introduction

Phylogeny is a very important field to understand evolution and the organization of life. However, many molecular phylogenies are built using sequences sampled from only a few genes. Besides, most methods depend highly on manual study and analysis, making the construction of phylogeny based on whole genomes difficult and error prone. The problem of analyzing genomes, however, is very similar to information retrieval from a large set of documents. In both problems, it is necessary to deal with an enormous amount of information, and to find semantic links between data. Fortunately, there are very good algorithms to deal with information retrieval. Singular value decomposition (SVD), for example, is used with great

Leandro S. Marcolino · Marcos A. dos Santos
Departamento de Ciência da Computação, Universidade Federal de Minas Gerais / UFMG
Av. Antonio Carlos 6627, Belo Horizonte, Minas Gerais, 31270-010, Brasil
e-mail: `soriano@dcc.ufmg.br, marcos@dcc.ufmg.br`

Bráulio R.G.M. Couto
Programa de Doutorado em Bioinformática, UFMG and Curso de Ciência da Computação,
Centro Universitário de Belo Horizonte / UNIBH
Av. Antonio Carlos 6627, Belo Horizonte, Minas Gerais, 31270-010, Brasil
e-mail: `bcouto@acad.unibh.br`

M.P. Rocha et al. (Eds.): IWPACBB 2010, AISC 74, pp. 225–232, 2010.
springerlink.com   © Springer-Verlag Berlin Heidelberg 2010

success (Berry *et al.* 1994). For example, linear algebra methods are used even by Google, enabling a better comprehension of a system as complex as the Internet (Eldén 2006; Stuart *et al.* 2002) presents a method to build phylogeny trees using SVD to analyze genomes. The method is demonstrated with vertebrate mitochondrial genomes, and is later used to analyze whole bacterial genomes and whole eukaryotic genomes (Stuart and Berry 2004). Linear algebra methods are also used to study the different genotypes in the human population (Huggins *et al.* 2007).

Visualization techniques are essential to better analyze complex systems and can be very helpful to categorize species. There are a number of visualization tools to study a single genome (Lewis *et al.* 2002; Engels *et al.* 2006; Rutherford *et al.* 2000; Stothard and Wishart 2005; Gibson and Smith 2003; Ghai *et al.* 2004). However it is desirable to visualize the relationships between a set of genomes, in order to better comprehend the species. In Xie and Schlick (2000) is presented a visualization technique using SVD to analyze chemical databases. In this paper, we used that technique as a basis to develop a method for using genomes to visualize relationships among species in space (2D and 3D). This can facilitate the construction of phylogeny trees, enabling the analyzer to quickly have insights in the similarities between the different species. We are going to show the results of our approach using 832 mitochondrial proteins obtained from 64 whole mitochondrial genomes of vertebrates.

2 Material and Methods

2.1 Sequence Data

We used the same set of proteins as Stuart *et al.* (2002), 64 whole mitochondrial genomes from the NCBI genome database, each one with 13 genes, totaling 832 proteins in the data set. The following species were used in this paper: *Alligator mississippiensis, Artibeus jamaicensis, Aythya americana, Balaenoptera musculus, Balaenoptera physalus, Bos taurus, Canis familiaris, Carassius auratus, Cavia porcellus, Ceratotherium simum, Chelonia mydas, Chrysemys picta, Ciconia boyciana, Ciconia ciconia, Corvus frugilegus, Crossostoma lacustre, Cyprinus carpio, Danio rerio, Dasypus novemcinctus, Didelphis virginiana, Dinodon semicarinatus, Equus asinus, Equus caballus, Erinaceus europaeus, Eumeces egregius, Falco peregrinus, Felis catus, Gadus morhua, Gallus gallus, Halichoerus grypus, Hippopotamus amphibius, Homo sapiens, Latimeria chalumnae, Loxodonta africana, Macropus robustus, Mus musculus, Mustelus manazo, Myoxus glis, Oncorhynchus mykiss, Ornithorhynchus anatinus, Orycteropus afer, Oryctolagus cuniculus, Ovis aries, Paralichthys olivaceus, Pelomedusa subrufa, Phoca vitulina, Polypterus ornatipinnis, Pongo pygmaeus abelii, Protopterus dolloi, Raja radiata, Rattus norvegicus, Rhea americana, Rhinoceros unicornis, Salmo salar, Salvelinus alpinus, Salvelinus fontinalis, Scyliorhinus canicula, Smithornis sharpei, Squalus acanthias, Struthio camelus, Sus scrofa, Sciurus vulgaris, Talpa europaea, and Vidua chalybeata.*

2.2 Representation Method

In order to visualize the genomes, we must represent each one as a point in space. The distance between the points should represent the differences in the genomes as a whole. Therefore, we might expect similar species to be close together in space. The genome proteins were represented as vectors of frequencies of groups of amino acids. In this paper, a sliding window of size 3 was used to measure the frequency. To represent the genome we used the vector sum of all its proteins. We are going to evaluate the appropriateness of this representation in the sequence. Therefore, we can obtain a database of genomes, S, as a rectangular matrix, X, where each line corresponds to one of the n genomes:

$$X = (X_1, X_2, \ldots X_n)^T = \begin{pmatrix} x_{11} & x_{12} & \cdots & x_{1m} \\ x_{211} & x_{22} & \cdots & x_{2m} \\ \vdots & \vdots & \ddots & \vdots \\ x_{n1} & x_{n2} & \cdots & x_{nm} \end{pmatrix}$$

As can be seen, the representation cannot be visualized in this high-dimensional space. With 20 amino acids, and considering that unknown amino acids are represented as a separated letter of the alphabet, each genome vector has m = 2^{13} = 9, 261 dimensions. Therefore, to generate a suitable visualization, it is necessary to reduce the dimensionality of the space, with the minimum loss of information. When a representation in reduced space, Y, is generated for the database matrix X, we can calculate an error function E as following:

$$E = \sum_i \sum_j \left(\delta_{ij} - \gamma_{ij} \right)^2$$

where δ_{ij} is the *euclidean distance* between genome i and j in the original space, represented in the matrix X, and γ_{ij} is the *euclidean distance* between genome i and j in the reduced space, represented in the matrix Y. The best representation of S in the reduced space will be the Y with the minimal associated error function. Therefore, we must solve an unconstrained optimization problem. Many methods can be used to solve this problem. In Xie and Schlick (2000), the truncated-newton minimization method is used. In this paper, we used a technique based on the interior-reflective Newton method. Singular value decomposition (SVD) is a popular method to reduce the dimensionality of a space, keeping the fundamental semantic association among the vectors in that space. Therefore, a good initial solution for the optimization problem can be obtained using the singular value decomposition (SVD) of X. The matrix is represented as X = $U\Sigma V^T$, where U = [u_1 u_2 $\ldots$ u_p], Σ = $diag(\sigma_1, \sigma_2, \ldots, \sigma_p)$, V = [$v_1$ $v_2 \ldots v_p$]. An approximation of X in reduced space (X_k) is given by:

$$X_k = \sum_{i=1}^{p} u_i \sigma_i v_i^T \, ; k \le p.$$

In this paper, we generated both two and three dimensional representations. We used a rank 2 approximation of X as the initial solution for the former, and a rank

3 approximation as the initial solution for the latter. After the optimization procedure, we have the best representation of the genomes to be visualized in a reduced
space.

3 Results and Discussion

We used the proposed approach to generate two and three dimensional visualiza-
tions of 64 whole mitochondrial genomes with 832 proteins. First, we are going to
evaluate if the *euclidean distance* of genomes using the chosen representation is
suitable to evaluate the similarities between them. Couto *et al.* (2007) showed that
the similarity of genome sequences can be measured by the *euclidean distance* in
a reduced dimensional space of tripeptides descriptors. They found a correlation
between the euclidean distance and global distance sequence alignment of +0.70.
To perform a similar analysis we created 64 supersequences by concatenating the
13 genes from each organism. These supersequences were compared by using
global edit distance between each pair of sequences and euclidean distance in the
high-dimensional space. As in Couto *et al.* (2007), the correlation between the edit
distance and *euclidean distance* was +0.70, but this time in a cubic model (P <
0.01; Figure 1). We can see, therefore, that the *euclidean distance* of genome sequences using the chosen representation can be used as a measure of similarity.

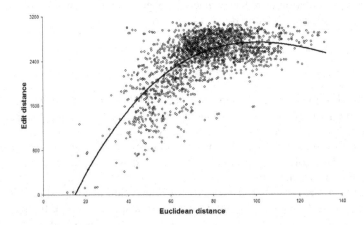

Fig. 1 Scatter plot of euclidean distance and global edit distance

We classified the species according to the class. Therefore, the following
groups were used: *Aves, Mammalia, Reptilia, Actinopterygii, Sarcopterygii, Chon-
drichtyes*. In Figure 2 we can see the 2D and 3D results. As can be observed, the
different class had a tendency to form groups in space. In the 2D graph we can see
that mammals (*mammalia*) are in the bottom, birds (*aves*) are in the upper
left, reptiles (*reptilia*) are generally in the middle left, and fishes (*actinopterygii,
sarcopterygii, chondirchthyes*) are in the upper right. It is notorious how the birds
are close together in a single cluster. In the results in 3D the classes are even better

clustered. This time, reptiles, birds and fishes are in distinctly separated groups. Only the class of the fishes are somewhat mixed.

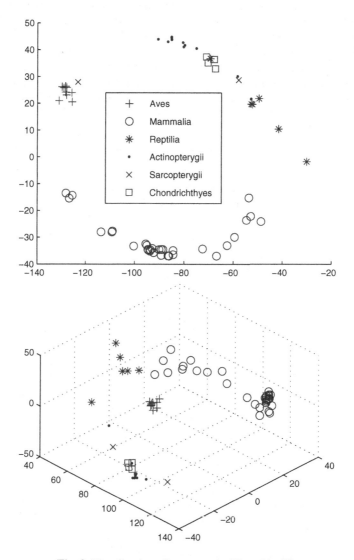

Fig. 2 Visualization of genomes in 2D and in 3D

It is interesting to observe the relationships between the classes, as similar groups tend to be near in space. The position of the class in the graphs seems to be related to the evolutionary scale. Considering the 2D graph as an ellipse, we can see that the reptiles are between the mammals and the fishes. In 3D this can be observed a second time. However, the evolutionary relationship between reptiles and birds is more clear in 3D, as there is no group between them.

Both in 2D and in 3D, mammals form a clearly distinct group from all other classes. They occupy a vast area, which might indicate more extensive diversity. We can also note that some mammals form clusters, what might be interesting to analyze. In order to better explore how the mammals are organized we separated this class in nine different groups: (i) *Prototheria*, corresponding to species in this subclass; (ii) *Marsupialia*, corresponding to species in this infraclass; (iii) *Chiroptera*, corresponding to species in this *ordo*; (iv) *Cetartiodactyla*, corresponding to species in this *superordo*; (v) *Carnivora*, corresponding to species in this *ordo*;

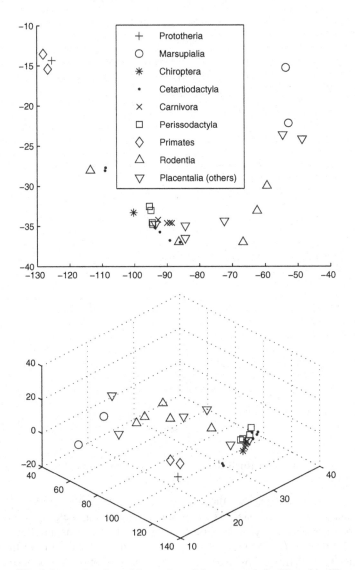

Fig. 3 Approximation of the region of the mammals in 2D and in 3D

(vi) *Perissodactyla*, corresponding to species in this *ordo*; (vii) *Primates*, corresponding to individuals in this *ordo*; (viii) *Rodentia*, corresponding to individuals in this *ordo*; (ix) *Placentalia*, corresponding to all other individuals that are in this infraclass, but were not classified in any other group. In Figure 3 we can see an approximation of the region of the mammals with this new classification. Similar species appeared close together, as was expected. This shows another advantage of the proposed method: as each genome is represented as points in space, we can easily select a region to better explore, zooming in and out in the graph as appropriate for the analysis.

The proposed method, however, allows another way to visualize a selected group of genomes. We can reduce the original set and run the algorithm a second time. Therefore, in order to better visualize the mammals, we executed the algorithm with only this class in the database. The result can be seen in Figure 4. It is interesting to note that the 2D graph has a similar elliptic format as in Figure 2. Clusters that were difficult to observe in Figure 3 are very clear in this graph. Similar species are again near to each other, showing visually the proximities of the genomes. In 3D the only group that mixed with the others is the Placentalia, but this was expected, as this group is very general, holding greatly different individuals. All other groups occupy distinct positions in space. We can see, therefore, that the proposed method allows many interesting observations and analysis of a group of genomes. Prespecified groups could be seen as clusters in the resulting graphs and the positions of the species seem to be related to their evolutionary stage. We also showed how approximating a region of the graph or running the algorithm a second time with a reduced data set allows a better insight of the relationships among selected groups of genomes. The resulting graphs can be generated both in two and in three dimensions for visualization.

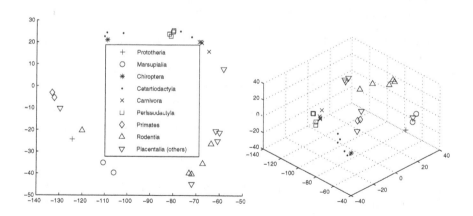

Fig. 4 Visualization of a reduced set in 2D and 3D

4 Conclusion

In this paper, we used a linear algebra method, followed by optimization, to visualize genomes in two and in three dimensional spaces. A set of complete mitochondrial genomes were used to test the algorithm. Graphs were generated to visualize the complete set and a reduced set of similar species. We noted that the method was able to automatically cluster some of the predefined groups and biologically similar species were represented as near points in space. We also noted that the position of the genomes in space seems to be related to the evolutionary stage of the species. Our future work is directed towards using this mechanism to visualize a large set of proteins. In this way, relationships between them can be easily observed and quickly explored, facilitating new discoveries. It would also be interesting to use this technique to explore a vast number of genomes, and further explore how it can be used to gain insight in evolution and in the phylogenetic relationships between the species.

References

Berry, M.W., Dumais, S.T., O'Brien, G.W.: Using linear algebra for intelligent information retrieval. Technical Report UT-CS-94-270, University of Tennessee, Knoxville, TN, USA (1994)

Couto, B.R.G.M., Ladeira, A.P., Dos Santos, M.A.: Application of latent semantic indexing (LSI) to evaluate the similarity of sets of sequences without multiples alignments character-by-character. Genetics and Molecular Research 6, 983–999 (2007)

Eldén, L.: Numerical linear algebra in data mining. Acta Numerica 15, 327–384 (2006)

Engels, R., Yu, T., Burge, C., et al.: Combo: a whole genome comparative browser. Bioinformatics 22(14), 1782–1783 (2006)

Ghai, R., Hain, T., Chakraborty, T.: Genomeviz: visualizing microbial genomes. BMC Bioinformatics 5, 198 (2004)

Gibson, R., Smith, D.R.: Genome visualization made fast and simple. Bioinformatics 19(11), 1449–1450 (2003)

Huggins, P., Pachter, L., Sturmfels, B.: Toward the human genotope. Bulletin of mathematical biology 69(8), 2723–2735 (2007)

Lewis, S., Searle, S., Harris, N., et al.: Apollo: a sequence annotation editor. Genome Biology (2002), doi:10.1186/gb-2002-3-12-research0082

Rutherford, K., Parkhill, J., Crook, J., et al.: Artemis: sequence visualization and annotation. Bioinformatics 16(10), 944–945 (2000)

Stothard, P., Wishart, D.S.: Circular genome visualization and exploration using cgview. Bioinformatics 21(4), 537–539 (2005)

Stuart, G.W., Berry, M.W.: An svd-based comparison of nine whole eukaryotic genomes supports a coelomate rather than ecdysozoan lineage. BMC Bioinformatics 5, 204 (2004)

Stuart, G.W., Moffett, K., Leader, J.J.: A comprehensive vertebrate phylogeny using vector representations of protein sequences from whole genomes. Molecular Biology and Evolution 19(4), 554–562 (2002)

Xie, D., Schlick, T.: Visualization of chemical databases using the singular value decomposition and truncated-Newton minimization. In: Floudas, C.A., Pardalos, P.M. (eds.) Optimization in Computational Chemistry and Molecular Biology: Local and Global Approaches, vol. 40. Kluwer Academic Publishers, Dororecht (2000)

A Hybrid Scheme to Solve the Protein Structure Prediction Problem

José C. Calvo, Julio Ortega, and Mancia Anguita

Abstract. This paper proposes an approach to the protein structure prediction (PSP) problem that inserts solutions provided by template-modeling procedures as individuals in the population of a multi-objective evolutionary algorithm. This way, our procedure represents a hybrid approach that takes advantage of previous knowledge about the known protein structures to improve the effectiveness of an *ab initio* procedure for the PSP problem. Moreover, the procedure benefits from a parallel and distributed implementation that allows faster and wider exploration of the conformation space. The experimental results obtained from the present implementation of our procedure show improvements with respect to previously proposed procedures in the proteins selected as benchmarks from the CASP8 set (up to 28% of RMSD improvement with respect to TASSER).

1 Introduction

Proteins are chains of amino acids whose sequence determines its 3D structure after a folding process. As the 3D structure of a protein exclusively determines its functionality [1] (transport and transduction of biological signals, the possible enzymatic activity of some proteins, etc.), there is a high interest in the determination of the structure of any given proteins. Experimental methods such as X-ray crystallography and nuclear magnetic resonance (NMR) allow the determination of the 3D structure of a protein although they are complex and expensive. Thus, only about the 25% of the known proteins has known structures [2]. The so called, protein structure prediction (PSP) problem is the approach to find the 3D structures of proteins by using computers.

The procedures for PSP can be classified into two main approaches, template modeling and template-free or *ab initio* procedures [3]. Template-based modeling

J.C. Calvo · J. Ortega · M. Anguita
Department of Computer Architecture and Computer Technology,
CITIC-UGR University of Granada
e-mail: jccalvo@ugr.es, jortega@ugr.es, manguita@ugr.es

M.P. Rocha et al. (Eds.): IWPACBB 2010, AISC 74, pp. 233–240, 2010.
springerlink.com   © Springer-Verlag Berlin Heidelberg 2010

are based on the experimental conclusion that homologous proteins have similar folds and functions. This way, once a homology between the query (newly sequenced) protein and some known protein is determined, it is possible to derive some knowledge about the structure and function of the query protein. The homology of two proteins is mainly determined by the similarity of their respective amino acid sequences. In fact, there is a rule of thumb [4] indicating that two proteins of about 100 amino acids and 25% of identities in their sequence are related by evolution with a probability of about 50%. Nevertheless, as there are also proteins with low levels of sequence similarity but similar structure and function, there are approaches not based on determining an alignment of their amino acids sequences but on other techniques such as profile methods, hidden Markov models, sequence signature libraries, etc.

The *ab initio* PSP procedures search the protein 3D structure without any knowledge of known structures. Thus, it does not require any homology and it can be applied when an amino acid sequence is not near to any other known one. Nevertheless, the high dimensionality of the space of conformations makes difficult for an *ab initio* procedure to find adequate structures for complex proteins, and nowadays it seems that template-based modeling is the most reliable and accurate approach to the protein structure prediction problem. Indeed, ab initio procedures frequently use searching procedures that start from some initial protein conformations that should represent an efficient sample of the protein 3D structure space to allow the search procedure to determine good solutions with high enough probability. Thus, for example, the Rosetta structure prediction algorithm [5] uses a two-step approach. In the first one, the protein is represented by using a low-resolution model based on the protein-backbone where some selected regions are modified taking into account a set of fragment libraries. Then, in the second step, the protein is explicitly represented by all its atoms and a specific energy function (the Rosetta energy function) is minimized by an optimization procedure.

In any case, the Protein Structure Prediction (PSP) problem remains unsolved due to the difficulties in the determination of an accurate energy function that makes it possible the identification of the native structure, and the complexity of the search process, a consequence of the large dimensionality of the protein conformation space and the high number of local minima in the energy landscape [6]. The different (PSP) procedures proposed up to now usually require many accesses to databases, and a lot of complex processing to determine a plausible protein conformation. Many alternatives have been considered to deal with the computing requirements of PSP. Among these, it is possible to enumerate the use of grid or global computing platforms including public resources [7] and parallel supercomputers [6]. In global computing platforms such as Rosetta@Home [8] and Predictor@Home [7] (based on BOINC [9]), the computing power is provided by heterogeneous system of computers interconnected by Internet, with different speeds, architectures, and operating systems.

In this paper we propose a hybrid template-based and *ab initio* procedure (Figure 1) that takes advantage of some common techniques in the template-based approaches to drive the search implemented by a multi-objective evolutionary algorithm towards the most promising zones of the space. This way, our multi-objective evolutionary algorithm offers the possibility not only to integrate different state of the art template-based approaches that provide plausible candidate structures (decoys) but also to take into account the influence of different knowledge-based potentials without the need to define a cost function that composes their effects. Moreover, the implicit parallelism shown by evolutionary algorithms can be considered as an interesting characteristic in order to take advantage of present high performance and heterogeneous computing platforms, as any of the conformation in the initial population can be determined in a different server. Once the initial population is built, a parallel platform could efficiently implement the multi-objective optimization procedure, to determine the best conformation.

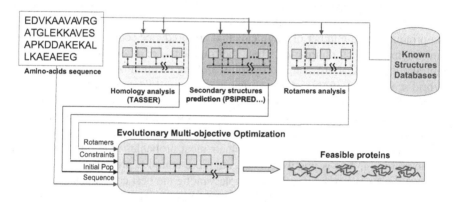

Fig. 1 A hybrid PSP process joining databases information and *ab initio* methods

In the paper, Section 2 introduces the knowledge based methods used in this hybrid scheme. The next section describes the concepts related with multi-objective optimization, and different methods to improve our multi-objective protein structure predictor by reducing the search complexity and the parallel implementations of the multi-objective approach to PSP problem. Finally, Section 4 provides the experimental results, and Section 5 the conclusions of the paper.

2 Knowledge-Based Methods

In this paper we use torsion angles to represent the conformation of the protein. Three torsion angles are required in the backbone and some additional torsion angles depending on the side-chain per each amino acid [1].

Although the PSP problem implies to predict the tertiary structure of a given protein from its primary structure, it could be a good idea to use predictions of the secondary and super-secondary structures as they give us information about the amino acids involved in one of these structures, determining some constraints in the torsion angles of each amino acid. In order to get the super-secondary structure given its secondary structure, we have to analyze the conformation of the residues in the short connecting peptide between two secondary structures. They are classified into five types, namely, a, b, e, l or t [10]. Sun *et al.* [10] developed a method to predict the eleven most frequently occurring super-secondary structures: H-b-H, H-t-H, H-bb-H, H-ll-E, E-aa-E, H-lbb-H, H-lba-E, E-aal-E, E-aaal-E and H-l-E where H and E are α helix and β strand, respectively. This way a reduction in the search space of the PSP problem is obtained.

Dunbrack et al.[11] give many rotamers libraries that help us to identify constraints about these torsion angles. These libraries have statistical information about side-chain torsion angles given the backbone torsion angles.

To be competitive, present *ab initio* methods should include strategies to start from good enough solutions or solutions that aid in the searching process. For example, as small proteins can be predicted easier than large ones (the conformation space grows exponentially with the number of amino acids) many procedures divide the proteins into a number of fragments that are predicted separately by searching into fragment structure libraries. Then, the fragments are assembled through different alternatives that are sampled by the searching or optimization procedure.The hybrid scheme here proposed uses different strategies to determine the initial population. The simpler one is to set each variable in the protein conformation to a random value (using the constraints in the corresponding variable). Other possibility is to use a probabilistic method that considers the rotamer libraries [11] to set the variables of each amino acid to their most probable value. Moreover, as it is shown in Fig. 1, it is possible to execute more complex procedures to take into account the known structures. For example, among the best current approaches for PSP are TASSER [12] and ROSETTA [5]. TASSER starts with a template identification process by iterative threading through the program PROSPECTOR_3, which is able to identify homologous and analogous templates. Then, the configuration is divided into continuous aligned fragments with more than five residues, and a Monte Carlo sampling procedure is applied to generate different assemblies of these protein fragments. Finally, the clustering program SPICKER is applied for model selection. ROSETTA also combines small fragments of residues (obtained from known proteins) by a Monte Carlo strategy. This way, in these procedures some kind of template-modeling is firstly applied before a random exploration of the conformation space spanned by different combining alternatives. The solutions provided by these procedures could be included in the initial population of an evolutionary optimization procedure to help in the search process as those solutions encapsulate the information about known structures.

3   The Proposed *ab initio* Multi-objective Approach

Multi-objective optimization [13] can be defined as the problem of finding a vector ($x = [x_1, x_2, ..., x_n]$) that satisfies a given restriction set ($g(x) \leq 0, h(x) = 0$) and optimizes the vector of objectives $f(x) = \{f_1(x), f_2(x), f_m(x)\}$. The objectives are usually in conflict between themselves, thus, optimizing one of them is carried out at the expense of the others. This leads to the need of making a compromise, which implies the concept of Pareto optimality. In a multi-objective optimization problem, a decision vector x* is said to be a Pareto optimal solution if there is not any other feasible decision vector, x, that improves one objective without worsening at least one of the other objectives.

In the last few years, some multi-objective approaches to the PSP problem have been suggested [1, 14]. For example, as indicated in [14], there are works that demonstrate that some evolutionary algorithms improve their effectiveness when they are applied to multi-objective algorithms. Indeed, in [1] it is argued that PSP can be naturally modeled as a multi-objective problem because the protein conformations could involve tradeoffs among different objectives as it is experimentally shown by analyzing the conflict between bonded and non-bonded energies. More arguments about the usefulness of multi-objective approaches to PSP based on effectiveness and problem simplification can be found in [15, 3].

Although a realistic measure of protein conformation quality should probably imply considering quantum mechanics principles, it would be too computationally complex to become useful. Thus, as it is usual, we have used the AMBER99 energy function; in particular, we have used these implemented in the TINKER library package. We propose a cost function with three objectives: the bond energy, the non-bond Van Der Waals energy and other for the rest of non-bond terms. A no formal proof about conflict between bond and non-bond can be found in [1]. Van Der Waals can hide the other non-bond terms because has higher change range, so it was separated from them. The algorithm also preserves the known structures with another objective that measures the similarity to the initial proteins.

We have also included two new strategies to improve the performance of an evolutionary algorithm solving the PSP problem: *(1) Simplified search space*: In the first part of the EA the search space is a simplification of the real one. This search space consists in only one variable, with only four possible values, per amino acid. This way, the EA can take into account the diversity of the search space. After this period, the search space becomes the real one. *(2) Amino acid mutation probability*: A new procedure to manage the mutation probabilities has been included. It takes into account that bond energies are independent of the the location of the corresponding amino acid, whereas the non-bonded energies depend on the present shape and structure as it is shown in Figure 2. Hence, this procedure tries to mutate with more probability the amino-acids that make better changes.

As many biological and medical applications, PSP requires intensive computation capabilities on large and usually distributed databases. It needs high performance computing capabilities, not only to reduce the time required to perform a single prediction, but also to get more accurate 3D protein conformations and to increase the size of the considered proteins. In [16] four paradigms of parallel multi-objective optimization evolutionary algorithms (master-worker, island, diffusion and hierarchical) are analyzed with respect to their migration, replacement and niching schemes, and generic parallel formulations for these parallel procedures are provided. We have implemented several parallel approaches for our multi-objective

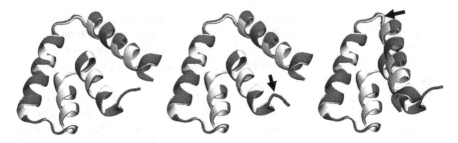

Fig. 2 Protein evolution: (left) initial protein, (center) changing a torsion angle in a final part and (right) changing a torsion angle in the center of the protein. It could be more interesting to generate big changes than little ones

PSP procedure. They are based on NSGA and PAES as it is described in [17]. The results in this paper corresponds to that obtained by our PAES version.

4 Experimental Results

In this section we analyze the results obtained with our parallel multi-objective algorithms. To evaluate the accuracy of the protein structures found we consider the RMSD [1] measure that gives the similarity (lower RMSD values are better) between the predicted and known native 3D structures. We also consider GDT-TS measure (higher GDT-TS values are better) that compute the percent of residues that can fit under distance cutoff 1, 2, 4 and 8 Å. We have run our algorithms by using a benchmark set that includes Free-Modeling proteins in CASP8 of different sizes and characteristics: T0397, T0416, T0496 and T0513. Initial population use TASSER results from CASP8, to avoid new knowledge in the Data Bases. Table 1 compares the results obtained with the TASSER [12] algorithm.

We have executed the algorithm along 250.000 cost function evaluations and we have selected the solution with better RMSD in the Pareto front. As it is shown in Table 1, the hybrid algorithm provides enough good solutions compared with TASSER in the free modeling proteins, due to its *ab initio* behaviour when no much knowledge can be extracted form the sequence.

Table 1 Provided algorithm (Hybrid PAES, HPAES) versus TASSER approach

Protein	# amino-acids	HPAES RMSD	TASSER RMSD	HPAES GDT-TS	TASSER GDT-TS
T0397	82	10.981 Å	11.239 Å	28.35	28.96
T0416	52	9.407 Å	12.934 Å	43.27	41.23
T0496	120	11.965 Å	11.885 Å	23.96	23.96
T0513	69	4.292 Å	4.297 Å	67.03	67.03

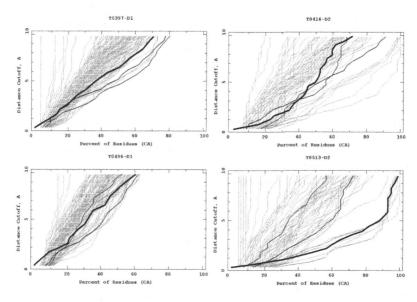

Fig. 3 Comparative with CASP8 algorithms in T0397, T0416, T0496 and T0513 proteins respectively. GDT analysis: largest set of CA atoms (percent of the modeled structure) that can fit under DISTANCE cutoff: $0.5Å$, $1.0Å$,..., $10.0Å$. Our algorithm is represented by the thicker line. Other three of the best procedures for T0397 have been selected to show their relative performance in different proteins

As it is shown in the curves of Figure 3, the relative performance of a given procedure can change for different proteins. This way, procedures that outperform ours in T0397 are outperformed in other proteins. As it can be seen in the four proteins considered, the procedure here proposed is among the best procedures.

5 Conclusions

The PSP problem joins biological and computational concepts. It requires accurate and tractable models of the conformations energy and there is still a long way to go to find the procedure that outperforms every other previous approach for all proteins. Our contribution in this paper deals with a new procedure for PSP based on hybrid multi-objective optimization that makes it possible to join the knowledge about protein structures (for example provided by template-based modeling) and *ab initio* algorithms. It includes strategies to reduce the search space, some heuristics to improve the quality of the solutions and an initial phase to get enough good solutions based on previous knowledge to initialize the optimization process. Moreover, our parallel implementation of PAES improves the computation efforts by using an adaptive probability of mutation in each amino acid. Particular issues for PSP will be addressed in future work. For instance, more research is required to benefit from the different meta-stable conformations included in the obtained Pareto front [15].

Acknowledgements. This paper has been supported by the Spanish Ministerio de Educacion y Ciencia under project TIN2007-60587. The hybrid schema has been developed with the collaboration of A. Zomaya and J. Taheri (The University of Sydney).

References

1. Cutello, V., Narcisi, G., Nicosia, G.: A multi-objetive evolutionary approach to the protein structure prediction problem. J. R. Soc. Interface 3, 139–151 (2006)
2. Zhang, Y., Skolnick, J.: The protein structure prediction problem could be solved using the current pdb library. Proc. Natl. Acad. Sci. USA 102, 1029–1034 (2005)
3. Handl, J., Kell, D., Knowles, J.: Multiobjective optimization in bioinformatics and computational biology. IEEE/ACM Transactions on Computational Biology and Bioinformatics (TCBB) 4(2), 279–292 (2007)
4. Rychlewski, L., Jaroszewski, L., Li, W., Godzik, A.: Comparison of sequence profiles. strategies for structural predictions using sequence information. Protein Science 9, 232–241 (2000)
5. Rohl, C., Strauss, C., Misura, K., Baker, D.: Protein structure prediction using rosetta. Methods in Enzymology 383, 66–93 (2004)
6. Raman, S., et al.: Advances in rosetta protein structure prediction on massivelly parallel systems. IBM J. Res. & Dev. 52(1/2) (2008)
7. Taufer, M., An, C., Kerstens, A., Brooks, C.: Predictor@home: A 'protein structure prediction supercomputer' based on global computing. IEEE Transactions on Parallel and Distributed Systems 17(8), 786–796 (2006)
8. Bradley, P., Misura, K., Baker, D.: Toward high-resolution de novo structure prediction for small proteins. Science 309, 1868–1871 (2005)
9. Anderson, D.: Boinc: A system for public-resource computing and storage. In: Proc. Fifth IEEE/ACM Int'l Workshop Grid Computing (2004)
10. Sun, Z., Jiang, B.: Patterns and conformations of commonly occurring supersecondary structures (basic motifs) in protein data bank. Journal of Protein Chemistry 15(7) (1996)
11. Dunbrack, R.: Rotamer libraries in the 21st century. Curr. Opin. Struct. Biol. 12, 431–440 (2002)
12. Wu, S., Skolnick, J., Zhang, Y.: Ab initio modelling of small proteins by iterative tasser simulations. BMC Biol. 5 (2007)
13. Deb, K., Pratap, A., Agarwal, S., Meyarivan, T.: A fast and elitist multiobjective genetic algorithm: Nsga-ii. IEEE Trans. on Evol. Comp. 6, 182–197 (2002)
14. Day, R., Zydallis, J., Lamont, G.: Solving the protein structure prediction problem through a multiobjective genetic algorithm. Nanotech 2, 32–35 (2002)
15. Tantar, A.-A., Melab, N., Talbi, E.-G., Parent, B., Horvath, D.: A parallel hybrid genetic algorithm for protein structure prediction on the computational grid. Future Generation Computer Systems 23, 398–409 (2007)
16. Veldhuizen, D.V., Zidallis, J., Lamont, G.: Considerations in engineering parallel multiobjective evolutionary algorithms. IEEE Transactions on Evolutionary Computation 7(2) (2003)
17. Calvo, J., Ortega, J., Anguita, M.: Comparison of parallel multi-objective approaches to protein structure prediction. The Journal of Supercomputing (2009)

Author Index